A Primer on Fourier Analysis for the Geosciences

Time-series analysis is used to identify and quantify periodic features in datasets and has many applications across the geosciences, from analysing weather data, to solid-Earth geophysical modelling. This intuitive introduction provides a practical 'how-to' guide to basic Fourier theory, with a particular focus on Earth system applications. The book starts with a discussion of statistical correlation, before introducing Fourier series and building to the Fast Fourier Transform (FFT) and related periodogram techniques. The theory is illustrated with numerous worked examples using R datasets, from Milankovitch orbital-forcing cycles to tidal harmonics and exoplanet orbital periods. These examples highlight the key concepts and encourage readers to investigate more advanced time-series techniques. It concludes with a consideration of statistical effect-size and significance. This useful book is ideal for graduate students and researchers in the Earth system sciences who are looking for an accessible introduction to time-series analysis.

DR ROBIN CROCKETT is Reader in Data Analysis in the Faculty of Arts, Science and Technology at the University of Northampton, UK. He is a member of the IMA and the IOP and holds Chartered Scientist Status. He specialises in investigating periodic, recurrent and anomalous features in data, and has led a highly successful short course on Fourier analysis at the European Geosciences Union General Assembly for many years.

T0331554

A Primer on Fourier Analysis
for the Geosciences

ROBIN CROCKETT

University of Northampton

CAMBRIDGE
UNIVERSITY PRESS

CAMBRIDGE
UNIVERSITY PRESS

University Printing House, Cambridge CB2 8BS, United Kingdom

One Liberty Plaza, 20th Floor, New York, NY 10006, USA

477 Williamstown Road, Port Melbourne, VIC 3207, Australia

314-321, 3rd Floor, Plot 3, Splendor Forum, Jasola District Centre, New Delhi - 110025, India

79 Anson Road, #06-04/06, Singapore 079906

Cambridge University Press is part of the University of Cambridge.

It furthers the University's mission by disseminating knowledge in the pursuit of education, learning and research at the highest international levels of excellence.

www.cambridge.org
Information on this title: www.cambridge.org/9781107142886
DOI: 10.1017/9781316543818

First published 2019

A catalogue record for this publication is available from the British Library

Library of Congress Cataloging in Publication data
Names: Crockett, R. G. M. (Robin G. M.), author.
Title: A primer on Fourier analysis for the geosciences / Robin Crockett
(University of Northampton).
Description: Cambridge : Cambridge University Press, [2019] | Includes
bibliographical references and index.
Identifiers: LCCN 2018040454 | ISBN 9781107142886 (hardback) |
ISBN 9781316600245 (pbk.)
Subjects: LCSH: Fourier analysis. | Earth sciences–Mathematics.
Classification: LCC QA403.5 .C76 2019 | DDC 515/.24330155–dc23
LC record available at https://lccn.loc.gov/2018040454

ISBN 978-1-107-14288-6 Hardback
ISBN 978-1-316-60024-5 Paperback

Additional resources for this publication at www.cambridge.org/fourier

For The Elf, The Moth and The Beagle.

Contents

Preface   *page* xiii
Acknowledgements   xiv

1   What Is Fourier Analysis?   1
1.1   What Does This Book Set Out to Do?   2
1.2   Choice of Software   4
1.3   Structure of the Book   5
      1.3.1   Examples and Exercises   6
1.4   What Previous Mathematics Do You Need?   6

2   Covariance-Based Approaches   9
2.1   Covariance and Correlation   9
      2.1.1   Interpreting the Correlation Coefficient   10
      2.1.2   Limitations   13
2.2   Lagged Correlation   13
      2.2.1   Same-Sense Cross-Correlation   15
      2.2.2   Opposite-Sense Cross-Correlation   18
2.3   Autocorrelation   20
2.4   Simple Linear Regression   23
2.5   Periodic Features: Correlation and Regression with Sinusoids   24
2.6   What Correlation and Regression 'See'   27
2.7   Summary   29

3   Fourier Series   31
3.1   What Are Fourier Series?   32
3.2   Orthogonality   32
      3.2.1   Clarification of Orthogonality   33
      3.2.2   Alternative Justification of Fourier Series   33
3.3   Fourier Series for General Time Function   34

3.3.1	Alternative Justification Revisited	35
3.3.2	Angular Frequency	36
3.4	Symmetry	36
3.4.1	Pure Odd- and Even-Symmetry	37
3.4.2	General Mixed Symmetry	39
3.4.3	Symmetry and Phase	41
3.5	General Properties of Fourier Series	41
3.5.1	Symmetry and Harmonic Composition	41
3.5.2	Linearity	41
3.6	Complex Fourier Series	42
3.6.1	Negative Frequencies	43
3.6.2	Real Data and Frequency Components	43
3.6.3	Complex Data	44
3.7	Summary	45
4	**Fourier Transforms**	**46**
4.1	Frequency Functions	46
4.2	The Fourier Transform	47
4.2.1	Fourier Transform Notation	48
4.2.2	Angular Frequency	49
4.2.3	Integrability	49
4.3	General Properties of the Fourier Transform	50
4.3.1	Symmetry	50
4.3.2	Linearity	50
4.4	Fourier Transforms of Sinusoids	51
4.4.1	The Dirac δ-Function	51
4.4.2	Fourier Transforms of Sines and Cosines	53
4.4.3	Fourier Transform of a General Sinusoid	53
4.4.4	From Continuous to Discrete Functions	55
4.5	The Discrete Fourier Transform	56
4.5.1	Derivation of the DFT from Fourier Series	56
4.6	The DFT as a Linear Transformation	59
4.6.1	The DFT Matrix and the Fast Fourier Transform	59
4.6.2	The Inverse DFT	61
4.6.3	Linearity of the DFT	61
4.7	The DFT Frequency Spectrum	62
4.8	Nyquist–Shannon Sampling Theorem	64
4.8.1	Resolving the Nyquist Frequency	65
4.9	Summary	66

5   Using the FFT to Identify Periodic Features in Time-Series   69
5.1   The Standard FFT   69
       5.1.1   Other Implementations of the FFT   70
5.2   Time-Series with Few Frequencies   70
       5.2.1   Preliminary Investigation   71
       5.2.2   Frequency Spectrum, FFT   71
       5.2.3   Amplitude Spectrum   73
       5.2.4   Power Spectrum   76
       5.2.5   Further Comments   77
5.3   Time-Series with Many Frequencies   78
       5.3.1   Initial Investigation   78
       5.3.2   Power Spectrum   79
       5.3.3   Further Comments   82
5.4   Time-Series with Trends   83
       5.4.1   Initial Investigation   83
       5.4.2   Detrending the Data   85
5.5   Time-Series with Noisy Spectra   87
       5.5.1   Power Spectrum   87
5.6   Summary   89

6   Constraints on the FFT   92
6.1   Minimum Resolvable Frequency and Low-Frequency Features   92
       6.1.1   Illustration: Low and Zero Frequency Features   92
       6.1.2   Windowing (Tapering)   94
6.2   Maximum Resolvable Frequency and Above-Nyquist Features   96
       6.2.1   Aliasing   97
       6.2.2   Illustration: Aliased Frequency   98
6.3   Resolving Intermediate Non-Harmonic Frequencies   99
       6.3.1   Linear Dependence of Intermediate Frequencies   100
       6.3.2   Illustration: Tidal Harmonics   101
       6.3.3   Sinusoidal Dummy Data   102
6.4   Re-Windowing Data, Padding   105
       6.4.1   Illustration: Tidal Harmonics   105
6.5   Resolving Adjacent Frequencies   107
       6.5.1   Illustration: Tidal Harmonics   108
6.6   Missing Data   109
6.7   Summary   110

7   Stationarity and Spectrograms   112
7.1   Stationarity   112
    7.1.1   Strong (Strict) Stationarity   112
    7.1.2   Weak Stationarity   113
7.2   Short-Time Fourier Transform   113
    7.2.1   Time and Frequency Resolutions of the STFT   114
    7.2.2   Uncertainty Principle   114
7.3   Spectrograms   115
    7.3.1   Overlapping Sections and Time–Frequency Resolution   115
    7.3.2   Spectrogram: Stationary Data   116
    7.3.3   Spectrogram: Non-Stationary Data   118
7.4   Summary   121

8   Noise in Time-Series   123
8.1   Noise Colour   123
    8.1.1   White Noise   124
    8.1.2   Red, Brownian and Pink Noise   124
    8.1.3   Blue and Violet Noise   125
8.2   Statistical Characterisations   125
    8.2.1   Autocorrelation and Autoregression   125
    8.2.2   Autoregressive Noise Models   126
8.3   Power–Law Relationships   129
8.4   Real Data with Red-Noise   132
    8.4.1   Autoregression and Autocorrelation   132
    8.4.2   Power Law   134
8.5   Summary   135

9   Periodograms and Significance   137
9.1   The Schuster, or Classical, Periodogram   138
    9.1.1   Vector Summation   141
    9.1.2   Probability and Significance   141
    9.1.3   Comparing the Schuster Periodogram to the DFT   143
9.2   Lomb–Scargle Periodogram   144
    9.2.1   Probability and Significance   145
    9.2.2   Effective Nyquist Frequency   146
    9.2.3   Example: Exoplanet Orbital Period   147
    9.2.4   Further Comments   150
9.3   FFT Spectra and Significance   150
    9.3.1   Power, Energy and Variance   151
    9.3.2   Proportion of Variance Explained   152

	9.3.3	Significance, p-value	154
	9.3.4	Real Distinct-Frequency Data	156
9.4	Summary		157

| *Appendix A* | **DFT Matrices and Symmetries** | 159 |
| *Appendix B* | **Simple Spectrogram Code** | 164 |

Further Reading and Online Resources	170
References	172
Index	174

Preface

This book evolved from the Fast Fourier Transform (FFT) and time-series short-courses I have given at the European Geosciences Union (EGU) General Assemblies over several years. Those short-courses were, in turn, an evolution from *ad hoc* sets of postgraduate tutorial materials to something more organised and integrated, *i.e.* from one-to-one response-to-questions delivery to one-to-many questions-anticipated delivery.

I did not know what to expect when I gave the first EGU short-course in 2009 – would anyone even turn up? Well, many more people attended than the allocated room could sensibly accommodate – and, needless to say, I did not anticipate all the questions. However, its success led to an invitation to give a similar short-course the following year – which again more than filled the bigger room allocated. Since then the course has rolled forward from year to year, each version maintaining the core FFT coverage but with varying details in response to suggestions and feedback from the previous year.

This book is a primer and so, by definition, cannot cover everything. It cannot even cover every type of geoscience time-series but it aims to give early-stage researchers, including research students, in the geosciences – and other sciences – a basic but robust grasp of the essential properties of the FFT in the context of time-series analysis. It focuses on that one specific task and, noting that it is intended for non-specialists to fill in the details that more advanced books on Fourier theory tend to pass over, endeavours to do this by presenting intuitive and illustrative theory and explanations rather than more abstract ones. As well as presenting core theory and practice for those wishing to use the FFT for time-series analysis, it goes a little deeper into the theory to give early-stage researchers sufficient depth and awareness of the limitations in order to critically evaluate other people's application and interpretation of the FFT.

I hope you find it useful both of itself and as a pointer and introduction to more advanced texts.

Acknowledgements

I am grateful to Emma Kiddle, Susan Francis and Laura Clark – my editors at Cambridge University Press – who picked-up on my European Geosciences Union short-courses and invited me to propose an extended book version thereof. I am also grateful to Zoë Pruce, Sarah Lambert and Cassi Roberts at Cambridge University Press to whom it fell to keep me focused when the consequences of sports injuries I acquired as a younger man meant I had to pretty much suspend the preparation for the better part of a year.

Going further back, I would like to thank Bruce Malamud who first invited me to give a short-course at the 2009 European Geosciences Union General Assembly. I would also like to thank Phil Picton and Frédéric Perrier who said 'just do it' and gave me valuable support and encouragement throughout. And ... everyone who has developed and disseminated Fourier theory before me.

1

What Is Fourier Analysis?

Fourier analysis can be defined as mathematical analysis based on the representation of a complicated waveform, *e.g.* a time-series, as the linear combination of a specific set of sinusoids. This started with Fourier series: "Fourier theory" is the general term used to describe the branch of mathematics which generalises and extends Fourier series beyond its original application.

Fourier series were initially developed in the early nineteenth century by, and named for, J-B J Fourier (1768–1830). The problem Fourier was trying to solve had nothing to do with periodic features *per se* but was a heat transfer problem in a metal plate. In brief, and skipping over quite a lot of important mathematical concepts that we will return to, Fourier could solve the same problem for a plate bounded by a sinusoid. From there, he reasoned that if he could exactly represent the boundary of the plate in question as a linear combination of sinusoids of different periods then the solution to the plate problem would be the linear combination of all the individual sinusoidal solutions. This was the first time that anyone had formalised the expansion of functions as trigonometrical series of sinusoids (*i.e.* sines and cosines).

Fourier's ideas did not gain immediate acceptance owing to their unorthodoxy and unfamiliarity and have been further developed and formalised since he first proposed them. However, Fourier's original insight that it is possible to use periodic functions to represent other functions carries forwards into general Fourier theory.

In essence, the Fourier transform provides us with a different way of viewing the world around us by transforming data from a sequential domain, such as time (and interval), to a wave domain, such as frequency (and phase). Sometimes, and particularly for periodic phenomena, the wave domain is easier to work with and more informative. All equal-interval sequential data that arise in real-life are (discrete) Fourier transformable and, if there are periodic features in those data, the Fourier transform will give us frequency and

1

phase information. There are other, more advanced and specialised techniques for doing this, but many are based on or related to the Fourier transform, so an intuitive understanding of the Fourier transform and the necessary mathematics gives a solid basis for understanding those more advanced techniques.

Use of the Fourier transform extends beyond its relatively straightforward application in the identification of periodic features in data, *e.g.* time-series. It has applications in technologies that are used everyday and on which we depend: in essence pretty much everything which involves transmission or recording of data, including audio and video files in the entertainment and communications industries, involves use of the Fourier transform, or something closely related, in some form. In addition, there are countless scientific and mathematical applications beyond time-series analysis.

I first encountered the Fourier transform and general Fourier theory in my undergraduate mathematics courses. Based on my experience since then, including much tutoring and advising of non-specialists, my opinion is that the Fourier transform is best understood intuitively, at least by non-specialists. The mathematics can be, initially at least, more than a little daunting, and it is that initial dauntingness that I am addressing in this book.

1.1 What Does This Book Set Out to Do?

This book, and the European Geosciences Union short-courses it evolved from, were prompted by questions along the following general lines:

- What is the FFT?
- What does the FFT do?
- Why do I get a column of N strange-looking numbers out when I put a column of N 'ordinary' numbers in?
- Why do I get complex numbers out when I put real numbers in?
- OK, I know what the Fourier transform does but how do I interpret what the FFT gives me?

So, this book tries to answer such questions and give the reader an understanding of the underpinning properties of the discrete Fourier transform (DFT), as generally implemented via a Fast Fourier Transform (FFT) algorithm, in the context of time-series analysis for periodic features. With regard to the two descriptors, *i.e.* DFT ('discrete') and FFT ('fast'), although not strictly identical in meaning the terms DFT and FFT are to a great extent interchangeable in practice. In the early chapters where I am developing the theory, I will refer to the DFT as the theory applies to the discrete nature of the data and the transform. In the latter chapters where I am referring to the use of the specific tool(s) coded in software, I will refer to the FFT.

This book is a primer. It is an introduction to Fourier and associated analysis and theory which underpins many of the concepts and techniques used in time-series analysis such as least-squares spectral analysis (LSSA), wavelets, singular-spectrum analysis (SSA) and empirical mode decomposition (EMD). As a primer, it is not a complete discussion of the subject and is not intended to be. Also, it is not intended to be read in isolation either from other (introductory) books on time-series analysis in the geosciences (and other sciences) or, indeed, other books on Fourier analysis and theory. Although it is explicitly aimed at geoscientists, and the examples and exercises are oriented towards the geosciences, it should be suitable for other scientists and researchers who want a very introductory text to the field.

As a book on Fourier analysis, it is a mathematics book but is one which focuses on the identification of periodic features in real data – real time-series – and estimating their effect-size and significance. These are aspects of Fourier theory which are important in time-series analysis but which are often passed over by the majority of books on Fourier theory, which emphasise more abstract aspects less relevant to time-series analysis and specialise in, for example, digital signal processing, audio and video processing and allied fields. Some of the content of this book will be familiar to people who have specialised in fields such as signal analysis but that same material can be far from familiar to non-specialists, including – broadly, in this context – early career geoscientists. Also, although effect-size and, by implication, statistical significance are touched on in some other books, these aspects are generally not the focus of those books and it is rare to find a systematic consideration consistent with statistical correlation and linear regression.

In keeping with these aims, while the mathematical reasoning, logic and flow of ideas and concepts are presented, proofs are not presented: readers with a need or desire for proofs are referred to books listed in the Bibliography (Further Reading) and others recommended to them. This is not to deprecate mathematical proof: proof is vital to the development of mathematics and mathematical proofs are often elegant and illuminating of themselves for those who view the world mathematically. However, that does not include everyone, and there will be some whose overriding and pressing need is to resolve immediate questions in their branch of science. That said, I encourage readers to refer to proofs when their time permits as they will quite likely find something that I cannot fully present or develop here, or something that gives them insights to problems that have been perplexing them.

1.2 Choice of Software

I decided to use R, the Free and Open-Source Software (FOSS) for statistical computing for this book. This is because R is open-source and freely available for Microsoft, Apple and the open-source Linux and BSD operating systems, making it a good choice because it means that I can work examples using one piece of software knowing that it is generally available to readers. Also, there are many specialist add-on library packages for R which contain algorithms/code for more advanced time-series analysis, *e.g.* there are packages for the Lomb–Scargle periodogram, wavelets, empirical mode decomposition, singular-spectrum analysis, amongst others.

We will use the included datasets in the datasets package, usually included with the R base package, and the add-on library packages astrodatR, astsa, EMD, RSEIS, TideHarmonics, TSA and tseries. We will, in addition, use the lomb package for its implementation of the Lomb–Scargle periodogram. When loading library packages for datasets, look at the analytical techniques included, and look at those in other time-series packages, too: one of those techniques might be suited for a dataset you are investigating. Also, the Bibliography (Online Resources) includes some URLs of websites that host datasets and some of those websites are the original sources of the datasets in the R library packages. In addition, it is possible to export datasets from R, *e.g.* in tab-delimited or comma-separated text files, for loading into other software.

I have deliberately kept my use of R code/commands in the book as basic as possible. This is for two reasons. First, to make worked examples as accessible as possible to readers who are more familiar with other software. Second, I avoid using R-unique ways of doing things so as to keep the commands as generic as possible for readers who wish to translate commands to other software. Although syntax varies between software packages, comparable software packages will have commands for performing the FFT, correlating and linearly regressing vectors of numbers, adding rows and columns to arrays and matrices, ordering arrays and matrices, plotting basic line and point graphs, and so on. For example, the basic forward and inverse FFT commands in R are, for data-vector z, fft(z, inverse = FALSE) and fft(z, inverse = TRUE) respectively: in Matlab and Octave (open-source, Matlab-like) the corresponding commands are fft(z) and ifft(z), and in Scilab (open-source) the corresponding commands are fft(z, −1) and fft(z, 1).

With regard to R and the worked examples, I am assuming that you have a working command of basic R syntax, *e.g.* downloading, installing and loading library files, loading data-frames, referencing rows and columns in data-frames and matrices, basic plotting and saving data-objects and workspaces. However, don't worry: there are plenty of books and online resources for R, not least the documentation available from the R-project website and mirrors. Commands for the more specialised operations such as cross- and auto-correlations and FFT are explained as they are introduced. Also, R has comprehensive help files, including help files on datasets. If you are an expert R-user, you will identify where you can make use of R-unique commands.

1.3 Structure of the Book

The book opens with a consideration and extension of statistical correlation and linear regression in Chapter 2. Correlation and regression should be familiar from mathematics and statistics courses in undergraduate science programmes, and so this gives an accessible starting-point for the rest of the book, but also a return point for statistical effect-size and significance in the final chapter. Also, covariance and correlation are closely related to the Fourier transform, although the majority of the theory is beyond the scope of this book, and to LSSA periodogram approaches in the final chapter. Chapters 3 and 4 cover most of the core Fourier and discrete Fourier theory, which I have aimed to present in an intuitively accessible, if not always fully rigorous, manner that is suited to the time-series context.

Chapter 5 consists of worked examples to demonstrate the use of the FFT, *e.g.* amplitude and power spectra, and also to illustrate a systematic approach to calibrating the frequencies and periods in the FFT output spectrum. Chapter 6 considers the limitations of the FFT, *i.e.* what the FFT cannot do in the context of identifying periodic features, as these are aspects of the FFT that can be overlooked. Chapter 6 also describes how to construct dummy time-series to help calibrate FFT spectra, padding/truncating time-series to 'tune' for frequencies of interest, and overviews work-arounds for time-series with missing data.

The last three chapters address specific issues that arise in the preceding chapters. Chapter 7 considers non-stationary data, *i.e.* essentially time-series with varying frequency composition, and introduces basic spectrogram approaches for time–frequency analysis. Chapter 8 presents some basic characterisations of noise as likely to occur in the FFT spectra of geoscience time-series. All real time-series are noisy, *i.e.* contain random fluctuations which are not part of any deterministic or systematic 'signal' content, and, indeed, an underlying noise spectrum in a time-series can provide a reference for evaluating the statistical effect-size of 'signal' content. This provides a starting point for Chapter 9 which considers basic periodograms, as examples of LSSA techniques, and statistical effect-size and significance, both of which also link directly back to Chapter 2.

1.3.1 Examples and Exercises

With regard to examples and exercises, there is no substitute for 'playing with data' relevant to an investigation that you have in hand or are about to embark on. I cannot hope to cover all eventualities and if you have some data to 'play with', and can adapt the approaches illustrated in the worked examples to those data, then doing that will be at least as useful as working through exercises that I have included. To that end, I have deliberately kept examples to a sensible minimum which illustrate (a) typical features of geoscience time-series that you might reasonably expect to encounter and (b) generally applicable systematic ways of investigating time-series, accompanied by end-of-chapter exercises to build on the worked examples. The worked examples are presented in box-outs in the text and if you work through those then you will obtain the results and plots discussed in the text. Some of the examples and exercises are developed and extended over successive chapters so you might find it helpful to save data objects and/or workspaces as R data-files as you work through, for subsequent reloading.

1.4 What Previous Mathematics Do You Need?

This is a book on Fourier analysis, albeit an introductory one that is highly focused on one specific task, *i.e.* the detection of periodic features in time-series. Even though its core aim is to present the basic mathematical framework of the Fourier transform in accessible intuitive terms, it has to assume a starting level in terms of mathematics and statistics that the reader has at the outset. Broadly, these are:

- Basic statistics. We will start with covariance, correlation and linear regression to set the scene for the Fourier theory and then return to these in order to investigate effect-size and significance, *i.e.* correlation coefficient (R), coefficient of determination (R^2) and statistical significance (p-value). We will also take some of these ideas forward to briefly consider periodograms, which are closely related to the discrete Fourier transform.
- Fourier series. Fourier series is the starting point for Fourier theory and an understanding of Fourier series sets a solid foundation. We start our consideration of Fourier theory with a statement of Fourier series and a

review of the key concepts and properties that carry forward into more
general Fourier theory, including the discrete Fourier transform.

- Circular (trigonometric) functions. Fourier series are explicitly dependent
  on the properties of sines and cosines as circular functions, and their
  derivatives and integrals, and their relationship with complex exponentials.
- Complex numbers. Complex numbers are crucial to the standard notation
  used for Fourier theory. Complex numbers are arguably not strictly
  necessary for basic Fourier theory but it would be much, much more
  difficult to describe, understand, use and develop without them. Complex
  numbers have the general two-part form $z = x + iy$ where i is the imaginary
  number, $i.e.$ that 'imaginary' number defined according to its property
  $i^2 = -1$, and x and y are the real and imaginary parts respectively, and both
  are real numbers. Some branches of engineering and the sciences use j for
  the imaginary number rather than i as used in this and many other
  books.
- Linear algebra. Linear algebra, $e.g.$ matrix-vector algebra, is not necessary
  for standard/continuous Fourier theory but the Fast Fourier Transform,
  which is what we all basically use when Fourier-transforming in practice
  using digital computers, is essentially an orthogonal linear transformation
  (linear mapping) from a time (or space) basis (frame of reference) to a
  frequency basis. Using concepts from linear algebra simplifies some of the
  material and illustrates the optimisations inherent in Fast Fourier Transform
  algorithms. Also, vector arithmetic is crucial to the classical (Schuster)
  periodogram.
- Mathematical notation. Mathematics books can be very notation-heavy.
  I have tried to have as light a touch as I consider useful but you will need a
  general familiarity with, for example, the use of lower-case and upper-case
  letters, Greek and Roman letters to represent mathematical constants,
  operators and variables, integral and summation symbols, use of
  superscripts/indices and subscripts.

These are all subjects and aspects of mathematics which are often included in
secondary/high school A-Level, Scottish Higher, International Baccalaureate
and equivalent mathematics syllabuses and, with the possible exception of
Fourier series, are often included in mathematics and statistics courses taken as
part of undergraduate degrees in science subjects. Don't worry if some of these
are unfamiliar or out-of-practice as there are very many textbooks and online
resources available at secondary/high school level. Also, many universities
have support units for mathematics and statistics, as well as mathematics and
statistics departments, which will be able to help.

Having outlined the sort of mathematics background that the book starts from, let us move to Chapter 2 and start with a review and extension of covariance-based techniques such as correlation and linear regression before moving on to Fourier theory. As well as being a starting point for the main subject matter of this book, variations of correlation and linear regression can provide a lot of information regarding periodic features in time-series in their own right.

2

Covariance-Based Approaches

We will begin with a review of statistical correlation and associated techniques such as simple linear regression, which are both covariance-based techniques. Most readers will have encountered these techniques in mathematics or statistics courses taken as part of their science programmes and they form an important foundation to the study of Fourier techniques. Looking a little further ahead, these covariance-based approaches are simple examples of what is more generally referred to as Least-Squares Spectral Analysis (LSSA) and lead directly to the classical (Schuster) and Lomb–Scargle periodograms which we will consider briefly in Chapter 9.

Covariance-based approaches are related to the Fourier transform. The Fourier transform can be regarded as a short-cut to an extended correlation process and correlation techniques such as auto- and cross-correlation can be expressed in terms of Fourier transform integrals (for the continuous, integral Fourier transform) and summations (for the discrete Fourier transform, DFT). Note that in this chapter we are implicitly assuming that the time-series we are considering are equal-interval and this is a requirement that we will formalise when we consider the DFT in Chapter 4. However, there is no formal requirement that this is the case for least-squares techniques, and periodogram approaches are useful alternatives to the DFT when analysing unequal-interval time-series for periodic features.

2.1 Covariance and Correlation

Covariance is the joint variation (joint variance) between two variables. For two N-element discrete variables $x(n) = \{x_n\}$ and $y(n) = \{y_n\}$ where $n = 0, 1, \ldots, N - 1$ is the index, the covariance is defined as

$$\text{cov}(x, y) = \frac{1}{N} \sum_{n=0}^{N-1} (x_n - \bar{x})(y_n - \bar{y}) \tag{2.1}$$

where $\bar{x}$ and $\bar{y}$ are the arithmetic means of $x(n)$ and $y(n)$ respectively.

Pearson product-moment ('standard') correlation is defined in terms of normalised covariance, *i.e.* covariance divided by the product of the standard deviations. Therefore, taking the covariance as defined in Equation 2.1, the Pearson correlation coefficient is

$$R = \frac{\text{cov}(x, y)}{\text{sd}(x) \cdot \text{sd}(y)} \tag{2.2}$$

where $\text{sd}(x)$ and $\text{sd}(y)$ are the standard deviations of $x(n)$ and $y(n)$ respectively, and $-1 \leq R \leq 1$.

Both covariance and correlation can provide a measure of the strength of an association between the two variables $x(n)$ and $y(n)$. However, it is not necessarily the case that a strong covariance or correlation indicates an association between the variables and a strong covariance or correlation can arise by chance between variables that have no actual association. Thus, it is important to remember that neither covariance nor correlation (nor regression) imply causality/causation between variables.

Lastly, note that we are using the $n = 0, 1, \ldots, N - 1$ indexing convention, rather than the possibly more familiar $n = 1, 2, \ldots, N$ convention, for consistency throughout the book. However, as we note in the next chapter, whilst this convention makes certain aspects of notation and algebraic manipulation more straightforward, it has no mathematical bearing and, in general, we could use any sequence of N consecutive integers.

2.1.1 Interpreting the Correlation Coefficient

At one extreme, a correlation coefficient $R = 0$ indicates zero correlation, *i.e.* the variables are uncorrelated, and if we plot $y(n)$ against $x(n)$ the points lie in a random scatter with an approximately circular envelope and we get no sense of a 'best straight line' through the scatter. At the other extreme, correlation coefficients $R = 1$ and $R = -1$ indicate perfect positive and perfect negative correlation respectively. If we have perfect correlation and plot $y(n)$ against $x(n)$ all the points lie on a straight line, sloping up from left to right for positive correlation and down from left to right for negative correlation. Between these extremes, as we move from zero correlation to perfect correlation, the envelope around the point-scatters moves from approximately circular to approximately elliptical, with a visible sense of a

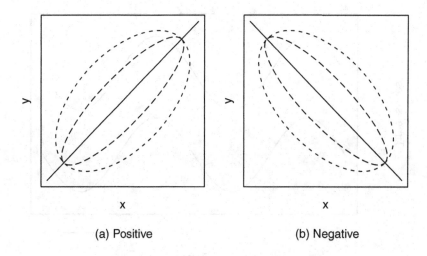

(a) Positive   (b) Negative

Figure 2.1 Correlation: schematic point-scatter envelopes.

long axis (major axis) sloping up from left to right for positive correlation and down from left to right for negative correlation. The narrower the ellipse, the better the correlation, limiting at a zero-width ellipse, *i.e.* a straight line along the long axis, for perfect correlation. This is shown schematically in Figure 2.1, where the the wider ellipses (dashed lines) indicate medium correlation relative to the narrower ellipses (long-dashed lines) which indicate high correlation and the straight lines which indicate perfect correlation.

If we are considering time-series, or other sequentially-indexed data, then (for time) we have $x(t) = \{x(t_n)\}$ and $y(t) = \{y(t_n)\}$, where $t = t_0, t_1, \ldots, t_{N-1}$ for $n = 0, 1, \ldots, N-1$. A more informative way of considering correlation in this general context is that if we had $x(t)$ and $y(t)$ with perfect positive correlation and we were to plot both $x(t)$ and $y(t)$ against t, *i.e.* against their common index, we would see that both vary exactly in step, in proportion and in the same sense at all points, as indicated in Figure 2.2. Similarly, if we had $x(t)$ and $y(t)$ with perfect negative correlation and we were to plot both $x(t)$ and $y(t)$ against t, we would see that both vary exactly in step, in proportion and in opposite senses at all points, as indicated in Figure 2.3. Also, if we had $x(t)$ and $y(t)$ with no correlation and we were to plot both $x(t)$ and $y(t)$ against t, we would see no 'in-steppedness', no proportionality, and approximately equal amounts of short-term same-sense and opposite-sense covariation.

More generally, if we had a degree of positive correlation and we were to plot both $x(t)$ and $y(t)$ against t, we would see that both vary generally

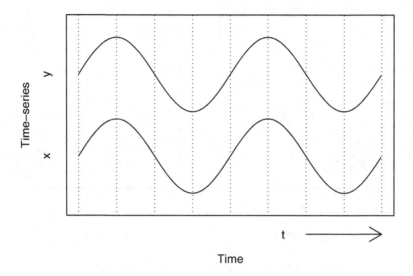

Figure 2.2 Perfect positive correlation (in-phase sinusoids).

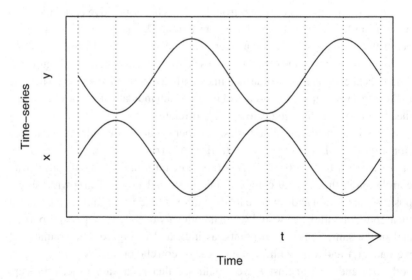

Figure 2.3 Perfect negative correlation (anti-phase sinusoids).

in step, approximately in proportion and generally in the same sense at most points. Similarly, if we had a degree of negative correlation and we were to plot both $x(t)$ and $y(t)$ against t, we would see that both vary generally in step, approximately in proportion and generally in opposite senses at most points.

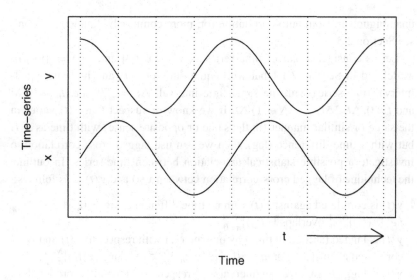

Figure 2.4 Zero correlation (quarter-phase sinusoids).

2.1.2 Limitations

This 'standard' correlation is fine for time-independent data but we generally need something more advanced when considering time-series (and other sequentially-indexed data). For example, consider the situation illustrated in Figure 2.4 where we have two identical sinusoidal waveforms $x(t)$ and $y(t)$ exactly one quarter-cycle out-of-phase. We might expect these to correlate strongly but, in fact, their correlation coefficient is zero (specifically, in this case, because sine and cosine are orthogonal, linearly independent, uncorrelated, and we say more about this in following chapters). In some circumstances, it might be that there is no association between $x(t)$ and $y(t)$ but, more generally, if we were to plot a pair of time-series and see something broadly similar to Figure 2.4, *i.e.* similar features in both time-series but separated by a time difference, we would expect there to be a statistical association.

2.2 Lagged Correlation

This is a technique we can use to investigate a possible relationship between a pair of time-series such as the simple case shown in Figure 2.4. In this context, correlation is generally referred to as cross-correlation to distinguish it unambiguously from auto-correlation (or, more commonly, 'autocorrelation', see Section 2.3).

Let us consider a pair of time-series $x(t) = \{x(t_n)\}$ and $y(t) = \{y(t_n)\}$, as defined in Section 2.1.1 but with equal durations, T, and both are equal-interval time-series with the same time interval, $\Delta t = T/N$, i.e. $t_n = n\Delta t$ and $t = 0, \Delta t, 2\Delta t, \ldots, (N-1)\Delta t$. If we know or suspect that $y(t)$ varies in the same or similar manner (in the same or opposite sense) with time as $x(t)$ but with a time difference (lag), then we can use lagged cross-correlation to investigate a possible statistical association between time-series. In outline, the technique of lagged cross-correlation between $x(t)$ and $y(t)$ is as follows:

- $y(t)$ is correlated against $x(t)$ with no time difference, lag $= 0$, i.e. $\{y(t_n)\}_{n=0}^{N-1}$ overlaps $\{x(t_n)\}_{n=0}^{N-1}$;
- $y(t)$ is lagged (shifted in time) by one interval with respect to $x(t)$ and is correlated against $x(t)$ at lag $= 1$, i.e. $\{y(t_n)\}_{n=0}^{N-2}$ overlaps $\{x(t_n)\}_{n=1}^{N-1}$;
- $y(t)$ is lagged again by one interval with respect to $x(t)$ and is correlated against $x(t)$ at lag $= 2$, i.e. $\{y(t_n)\}_{n=0}^{N-3}$ overlaps $\{x(t_n)\}_{n=2}^{N-1}$;
- and so on for lag $= 3, 4, \ldots,$ lag_max (as defined in the following paragraph);
- and also, in general, for 'negative' lags, i.e. lagging $y(t)$ in the opposite time-sense with respect to $x(t)$, i.e. $\{y(t_n)\}_{n=1}^{N-1}$ overlaps $\{x(t_n)\}_{n=0}^{N-2}$ at lag $= -1$, and so on for lag $= -2, -3, \ldots, -$lag_max.

In principle, we could lag $y(t)$ with respect to $x(t)$ up to maximum lags of $\pm(N-2)$ but, at that point, we would be correlating two points at one end of $y(t)$ against two points at the opposite end of $x(t)$, which is unlikely to be informative. In general, therefore, it is sensible to regard the maximum useful lag as being the biggest lag at which at least half of $y(t)$ overlaps with at least half of $x(t)$, i.e. lag_max $= N/2$ for N-even and lag_max $= (N-1)/2$ for N-odd.

Thus, in its most basic form, for two N-element equal-interval time-series $x(t) = \{x(t_n)\}$ and $y(t) = \{y(t_n)\}$ and for lag k with $-$lag_max $\leq k \leq$ lag_max, the basic lagged cross-correlation coefficient is

$$R(k) = \frac{\text{cov}(x_{k'}, y_k)}{\text{sd}(x_{k'})\,\text{sd}(y_k)} \tag{2.3}$$

where the covariance and standard deviations apply to the segments $x_{k'}$ and y_k that overlap at lag k and $-1 \leq R(k) \leq 1$.

This results in a sequence of values for the correlation coefficient, $R(k)$, at successive values of the lag, k. For $x(t)$ and $y(t)$ varying in the same sense, where $R(k)$ is a:

- local maximum, the time lag k is where lagged $y(t)$ is most in same phase/step with unlagged $x(t)$;
- local minimum, the time lag k is where lagged $y(t)$ is most in opposite phase/step with unlagged $x(t)$.

Interpreting the sequence of values for $R(k)$ with caution, for $x(t)$ and $y(t)$ varying in the same sense, a sequence with:

- recurrent maxima in $R(k)$ can indicate recurrent features in common between $x(t)$ and $y(t)$, with the recurrence intervals corresponding to the differences between values of k at successive local maxima in $R(k)$;
- periodic maxima in $R(k)$ can indicate periodic features in common between $x(t)$ and $y(t)$, with the period given by the constant difference between values of k at successive local maxima in $R(k)$.

For $x(t)$ and $y(t)$ varying in opposite senses, the same reasoning applies but we consider local minima (for peak negative correlation) rather than local maxima (for peak positive correlation).

Note that the value of $R(0)$ depends on $x(t)$ and $y(t)$ and, in general, $R(0) \neq 1$. Also in general, $R(k) \neq R(-k)$ for lag $\pm k$.

There are variants of the this basic form of lagged cross-correlation but all seek to correlate two time-series (or other sequentially-indexed data) with one shifted in time (or other sequential index) with respect to the other in order to investigate for similar features but offset in time (or other index).

2.2.1 Same-Sense Cross-Correlation

This example uses two built-in datasets, sunspot.year in the R datasets package which is annual sunspot totals over the period 1700–1988, and gtemp in the astsa library package, which is annual global mean land-ocean temperature deviations for the period 1880–2009. First, we will plot the two time-series: both are stored in R's time-series format but are basically vectors of sequential (annual) data with time information embedded as a header. This format makes plotting easy, as R interprets axis ranges from the embedded time information and automatically gives a line-graph. Thus we can plot the time-series, with basic axis labels, as shown in Figures 2.5 and 2.6 respectively, as follows.

```
plot(sunspot.year, xlab="Year", ylab="Sunspots")
plot(gtemp, xlab="Year", ylab="Temperature deviations")
```

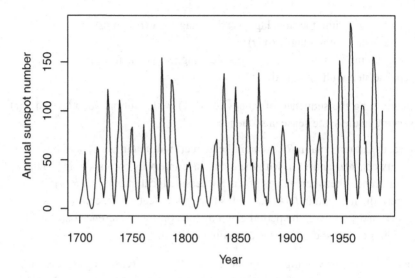

Figure 2.5 The sunspot.year time-series.

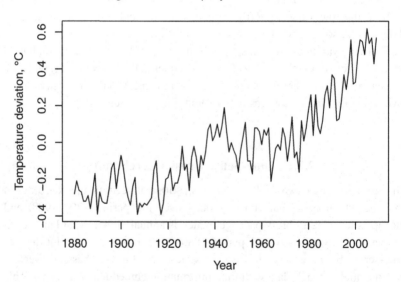

Figure 2.6 The gtemp time-series.

By inspection, we can see a *ca.* 11-year cycle in the sunspot numbers, *i.e.* sunspot numbers vary according to the *ca.* 11-year solar cycle, and a 'noisier' cycle in the land-ocean temperature deviations superimposed on an increasing trend.

To perform the cross-correlation, we first need to extract sunspot and temperature data over the common time interval, *i.e.* 1880–1988, using the window() command and then cross-correlate as follows using the ccf() command. Note that the as.numeric() command is not strictly necessary here but it strips all the embedded time-unit information resulting in two straightforward numeric vectors, and so calibrates the horizontal axis in lag-intervals rather than the embedded time-units.

```
sy2 <- as.numeric(window(sunspot.year, start=1880,
end=1988))
gt2 <- as.numeric(window(gtemp, start=1880, end=1988))
ccf(sy2, gt2)
```

This produces a basic correlogram plot with default labels. To produce the same correlogram but with more informative axis labels and without a default plot title (which is not needed here), as shown in Figure 2.7, we can use an extended form of the ccf() command as follows.

```
ccf(sy2, gt2, xlab="Lag, years",
ylab="Cross-correlation", main="")
```

Figure 2.7 shows that there is a weak statistical correlation which peaks for a lag of temperature behind sunspots of *ca.* 2 years. Although weak, it peaks

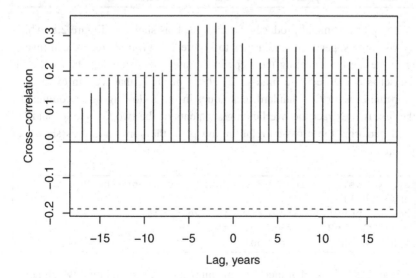

Figure 2.7 The gtemp – sunspot.year cross-correlation.

above the horizontal dashed line which shows the confidence interval, which defaults to 95%. Note that this does not of itself tell us that there is a causal lag of temperature behind solar activity: at most, it tells us that if there is a causal lag, then it is 2 years for these particular datasets. However, for example, White *et al.* (1997) identify similar lags for some similar datasets. Also note that for identification of lags in recurrent or periodic features, it can be appropriate to de-trend the data (see Section 5.4).

2.2.2 Opposite-Sense Cross-Correlation

The previous example demonstrated the form of the output we might expect to see when cross-correlating variables that vary in the same sense. To illustrate the form of the output we might expect to see when cross-correlating variables that vary in opposite senses, we will consider the solar irradiance proxy ^{10}Be. This varies in the opposite sense to sunspot activity, *i.e.* as solar irradiance increases the amount of ^{10}Be in the atmosphere decreases, and so the amount of ^{10}Be deposited in sediments decreases.

We can investigate this by repeating the steps in the previous example, using the beryllium time-series in the EMD library package. First, we can plot the data as shown in Figure 2.8, as follows.

```
plot(beryllium$year, beryllium$be, type="l",
xlab="Year", ylab="Beryllium Concentration")
```

The plot command produces the line-graph as shown in Figure 2.8, which shows some superficial similarities to Figure 2.5. From there, we can investigate the lag by repeating the steps in the previous example. The beryllium dataset is not stored as a time-series object meaning that we can extract the year-range using R's standard technique. In the following three commands, the first two extract the overlapping segments of the two time-series and the third performs the cross-correlation and plots the correlogram as shown in Figure 2.9.

```
sy2 <- as.numeric(window(sunspot.year, start=1700,
end=1985))
be2 <- beryllium$be[beryllium$year>=1700]
ccf(sy2, be2, xlab="Lag, years",
ylab="Cross-correlation", main="")
```

Figure 2.9 shows that there is a medium statistical correlation which peaks negatively for a lag of ^{10}Be behind sunspots of *ca.* 1 year. Note that this does

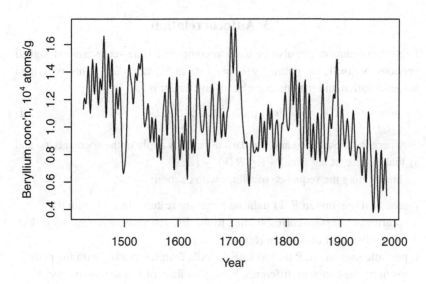

Figure 2.8 The beryllium dataset.

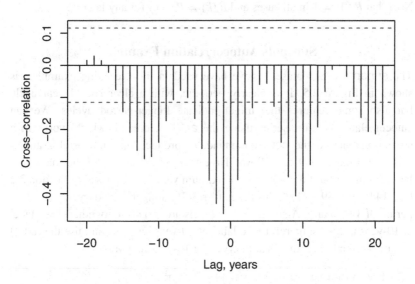

Figure 2.9 The beryllium – sunspot.year cross-correlation.

not of itself tell us that there is a causal lag of [10]Be behind solar activity: at most, it tells us that if there is a causal lag, then it is 1 year for these particular datasets.

2.3 Autocorrelation

Lagged correlation can also be used to compare a time-series against lagged versions of itself, by putting $y(t) = x(t)$ in Equation 2.3. This is termed autocorrelation and the autocorrelation coefficient is

$$R(k) = \frac{\text{cov}(x_{k'}, x_k)}{\text{sd}(x_{k'}) \cdot \text{sd}(x_k)} \tag{2.4}$$

where the covariance and standard deviations apply to the segments $x_{k'}$ and x_k that overlap at lag k and $-1 \leq R(k) \leq 1$.

Interpreting the sequence of values with caution:

- recurrent maxima in $R(k)$ indicate recurrent features in $x(t)$, with the recurrence intervals corresponding to the differences between values of k at successive local maxima in $R(k)$;
- periodic maxima in $R(k)$ indicate periodic features in $x(t)$, with the period given by the constant difference between values of k at successive local maxima in $R(k)$.

Note that $R(0) = 1$ in all cases and $R(k) = R(-k)$ for any lag $\pm k$.

Sunspots Autocorrelation Example

The sunspot.year dataset, as we have used in the preceding examples, is shown in Figure 2.5. It is clearly periodic, with a clear *ca.* 11-year cycle but also some features that might indicate longer-period cycles. We can autocorrelate this time-series using the acf() command, which works in a similar manner to the ccf() command, to produce the autocorrelogram as shown in Figure 2.10, as follows. Note that we are adjusting the maximum lag from the in-built default to the maximum value as discussed in Section 2.2 (*i.e.* $144 = (289 - 1)/2$) in order to resolve longer-period cycles (up to a period of 144 years). Also, the x-axis labels are too close together if we label at 10-year intervals, therefore we label at 20-year intervals and use the axis() command to add tick-marks at intervening 10-year intervals.

```
acf(as.numeric(sunspot.year), lag.max=144,
xaxp=c(0,140,7), xlab="Lag, years",
ylab="Autocorrelation", main="")
axis(1, at=c(10, 30, 50, 70, 90, 110, 130),
labels=FALSE)
```

Figure 2.10 shows a clear *ca.* 11-year cycle as expected. However, as shown by the decrease in the magnitudes of the periodic local maxima (and minima)

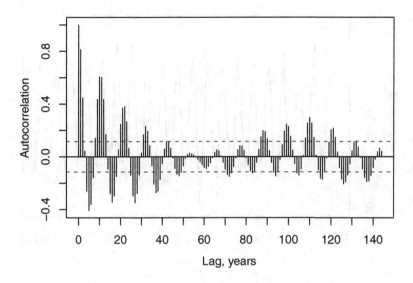

Figure 2.10 The sunspot.year autocorrelation.

to a minimum at *ca*. 5-6 11-year cycles followed by an increase to a maximum at *ca*. 10 11-year cycles, Figure 2.10 also shows evidence of a longer, *ca*. 110-year tenth sub-harmonic cycle (*i.e.* frequency one-tenth of the basic 11-year cycle).

Nottingham Temperature Autocorrelation Example

The nottem dataset included in the R datasets package is a 20-year (1920–1939) record of monthly mean temperatures in Nottingham, UK, in degrees Fahrenheit. This is a convenient dataset to illustrate autocorrelation and also to set the scene for the DFT in the next chapter, and is a dataset to which we will return to illustrate other techniques. This dataset is also stored as a time-series object in R and we can plot the time-series for an initial investigation, as shown in Figure 2.11, as follows.

```
plot(nottem, xlab="Year", ylab=expression("Temperature,
" * degree * F))
```

Figure 2.11 shows a clear annual cycle which appears to be approximately sinusoidal, noting some year-to-year variations in summer maxima and winter minima. It is much more uniform from cycle to cycle than the sunspots data and we can confirm this by producing the autocorrelogram, as shown

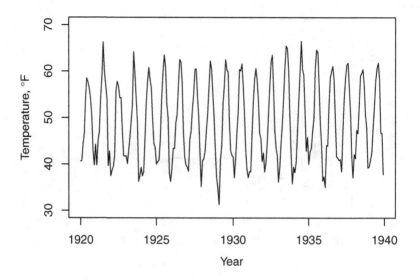

Figure 2.11 The nottem time-series.

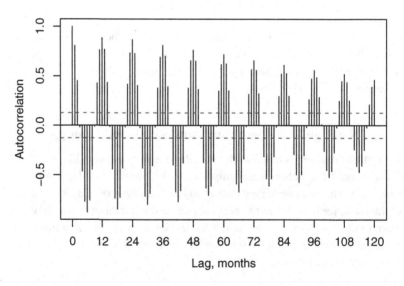

Figure 2.12 The nottem autocorrelation.

in Figure 2.12, as follows. Note that as with the previous example, we are adjusting the maximum lag from the in-built default to the maximum value (*i.e.* $120 = 240/2$) in order to resolve any longer-period cycles (up to a period of 120 months, 10 years).

```
acf(as.numeric(nottem), lag.max=120, xaxp=c(0,120,10),
xlab="Lag, months", ylab="Autocorrelation", main="")
```

As shown in Figure 2.12, the variation in the autocorrelation coefficient is highly periodic, as shown by the regular local maxima and minima at 1-year (12-month) lags with only a gradual decrease in magnitude of those maxima and minima as lag increases, corresponding to the slowly increasing mismatch of year-to-year variations as the lag increases. Also, and unlike the sunspots data, there are no longer-period variations, indicating that for this 20-year period there are no longer-period variations with periods up to 10 years.

2.4 Simple Linear Regression

As outlined in the preceding sections, statistical correlation measures the linearity of a relationship between two variables, *i.e.* the extent to which they are proportional to each other, but it does not directly measure the constant of proportionality. In order to do that, we need to use simple linear regression, a closely related technique.

If we have a pair of N-element variables $x(n) = \{x_n\}$ and $y(n) = \{y_n\}$, as for cross-correlation, then simple linear regression fits the model

$$y(n) = \alpha + \beta x(n) + \varepsilon(n) = \hat{y}(n) + \varepsilon(n) \tag{2.5}$$

according to best straight line that goes through the data, *i.e.*

$$\hat{y}(n) = \alpha + \beta x(n) \tag{2.6}$$

where α and β are the y-axis intercept and gradient of the regression line respectively and $\hat{y}(n) = \{\hat{y}_n\}$ is the set of regression estimates for $y(n)$. The N-element variable $\varepsilon(n) = \{\varepsilon_n\}$ is the set of residuals (*i.e.* errors) between the actual values, $y(n)$, and the regression estimates, $\hat{y}(n)$. The best straight line, *i.e.* the regression line, is the straight line that yields the smallest sum of the squared residuals.

The formulae for α and β are

$$\beta = \frac{\text{cov}(x, y)}{\text{var}(x)} \tag{2.7}$$

$$\alpha = \bar{y} - \beta\bar{x} \tag{2.8}$$

where var (x) is the variance of $x(n)$ and $\bar{x}$ and $\bar{y}$ are the arithmetic means of $x(n)$ and $y(n)$ respectively.

Thus, unlike correlation where there is no sense of explanatory (independent, conventionally x, horizontal) and explained (dependent, conventionally y, vertical) variables, in simple linear regression there is very clear explanatory-explained precedence in the variables. In other words, regressing $y(n)$ against $x(n)$ (*i.e.* $y(n)$ as a function of $x(n)$), as in Equations 2.5–2.8, is not the same as regressing $x(n)$ against $y(n)$ (*i.e.* $x(n)$ as a function of $y(n)$).

In simple linear regression, the goodness-of-fit is generally expressed by the coefficient of determination, R^2 ('R-squared'). This is the proportion of variance explained, *i.e.* the proportion of the variance of the explained variable that is explained by the explanatory variable(s). However, because the proportion of unexplained variance, *i.e.* the proportion of variance accounted for by the residuals, is easier to calculate in general (and particularly for linear regressions with more than one explanatory variable), the formula for R^2 is

$$R^2 = 1 - \frac{\text{var}(\varepsilon)}{\text{var}(y)} = 1 - \frac{rss}{tss} \tag{2.9}$$

where *rss* is the sum of the squared residuals and *tss* is the total sum of squares, *i.e.*

$$rss = \sum_{n=0}^{N-1} \varepsilon_n^2 = \sum_{n=0}^{N-1} (y_n - \hat{y}_n)^2$$

$$tss = \sum_{n=0}^{N-1} (y_n - \bar{y})^2 \tag{2.10}$$

For simple linear regression, *i.e.* two variables (one dependent on the other), the coefficient of determination (R^2) is equal to the square of the correlation coefficient (R) of the same two variables. This not the case for linear regression with two or more explanatory variables, *i.e.* the coefficient of determination from the linear regression of y against $x_A, x_B, \ldots$ is not generally equal to the sum of the squared correlation coefficients $R^2_{x_A y} + R^2_{x_B y} + \ldots$ because, in general, the variables $x_A, x_B, \ldots$ are not linearly independent (not completely uncorrelated).

We will make use of R^2 as the proportion of variance explained in Chapter 9 where we consider statistical effect-size and significance of DFT frequency components.

2.5 Periodic Features: Correlation and Regression with Sinusoids

We could, at least in principle, use correlation or simple linear regression to investigate possible statistical dependence of a time-series, $y(t)$, on a reference time-series, $x(t)$, which contains a periodic waveform or other feature that we know or suspect exists in $y(t)$. Clearly, there is an infinite number of possible general reference time-series for any actual time-series under investigation but if we restrict ourselves to basic periodic waveforms, we can start to formalise an approach.

The most basic periodic function is a sinusoid and, if we wish to ensure orthogonality, we can further restrict our consideration to sinusoids which have an integer number of cycles over the whole duration, T, of the time-series under investigation. This approach results in a set of sinusoids with frequencies $f_1 = 1/T$, the fundamental, and harmonics $f_2 = 2/T, f_3 = 3/T$, general kth harmonic $f_k = k/T$, and so on, plus an average (or 'dc') level, with frequency $f_0 = 0/T$. This still does not account for phase, but we can resolve that complication by making use of the trigonometric identity that allows us to express any sinusoid of frequency f_k and arbitrary phase ϕ_k as a sum of $\sin(2\pi f_k t)$ and $\cos(2\pi f_k t)$ terms (which are orthogonal), i.e.

$$
\begin{aligned}
A\cos(2\pi f_k t - \phi_k) &= A\cos(\phi_k)\cos(2\pi f_k t) + A\sin(\phi_k)\sin(2\pi f_k t) \\
&= \alpha_k \cos(2\pi f_k t) + \beta_k \sin(2\pi f_k t) \\
&= \alpha_k \cos\left(\frac{2\pi}{T_k}t\right) + \beta_k \sin\left(\frac{2\pi}{T_k}t\right)
\end{aligned} \tag{2.11}
$$

where $\alpha_k = A\cos(\phi_k)$ and $\beta_k = A\sin(\phi_k)$ are the constant coefficients of the cosine and sine terms respectively and T_k is the period of the kth harmonic, $T_k = 1/f_k = T/k$.

This means that we would simply need reference unit-amplitude sine and cosine waveforms at each harmonic frequency, f_k, and could use either linear regression to obtain coefficients α_k and β_k and the proportion of variance in our time-series explained by the kth harmonic sinusoid (R_k^2), or correlation if we simply wish to confirm an association, and obtain correlation coefficient (R_k). Note that owing to the orthogonality (linear independence) of the $\sin(2\pi f_k t)$ and $\cos(2\pi f_k t)$ terms, we can directly add the individual $R_{\cos}^2$ and $R_{\sin}^2$ values to obtain the overall R_k^2 value for the kth harmonic sinusoid. Orthogonality of sines and cosines is fundamental to Fourier analysis and is discussed in the next chapter.

Nottingham Temperature Regression Example

We can usefully illustrate linear regression using the nottem dataset. This comprises a list of 240 sequential values (20 years, 12 monthly values per year). In order to generate the annual sine and cosine (i.e. $f_{20} = 1/12 = 20/240$, 1 cycle over 12 months, 20 cycles over the 240-month period of the data) we first need to generate a time-base, $t_b = 0, 1, \ldots, 239$, then the sine and cosine terms as follows.

```
tb <- 0:239
ann.cos <- cos(2*pi*(1/12)*tb)
ann.sin <- sin(2*pi*(1/12)*tb)
```

Note that, more generally, being able to generate sinusoids of specific frequencies can be useful for checking that we are correctly interpreting the frequency indexing of the DFT, and we return to this in Section 6.3.

Next we perform the linear regressions and print (to the R console) the summaries of the regression outputs which contain the information we need.

```
lm.cos <- lm(nottem ~ ann.cos)
lm.sin <- lm(nottem ~ ann.sin)
summary(lm.cos)
summary(lm.sin)
```

We do not reproduce the full regression outputs here but these show that (subscripting in terms of the number of cycles over the 20-year period):

- the mean level, given by the '(Intercept)' coefficient of both regressions, is 49.04 °F;
- for the cosine $\alpha_{20} = -11.47$ and $R^2_{\cos,20} = 0.8994$ (89.94%);
- for the sine $\beta_{20} = -1.39$ and $R^2_{\sin,20} = 0.0132$ (1.32%).

with significance (p-value) and other information also given.

Therefore, the annual sinusoidal model which best explains the variation in the nottem data is

$$\text{ann.cyc} = -11.47 \cos\left(2\pi \frac{1}{12} t\right) - 1.39 \sin\left(2\pi \frac{1}{12} t\right) + 49.04 \qquad (2.12)$$

which we can express as

$$\text{ann.cyc} = -11.56 \cos\left(2\pi \frac{1}{12} t - 0.1206\right) + 49.04 \qquad (2.13)$$

where the amplitude is $|A_{20}| = 11.56$, and the phase is $\phi_{20} = 0.1206$ which corresponds to a time-offset of 0.23 months relative to $t = 0$ (January 1920), i.e. the annual sinusoid reaches annual minimum and maximum values in January and July respectively. Also, we can add the individual $R^2_{\cos,20}$ and $R^2_{\sin,20}$ values to obtain the overall R^2_{20} value for the annual sinusoidal model, because the explanatory sine and cosine variables are orthogonal, and this is $R_{20}^2 = 0.9126$ (91.26%).

We can go a little further and construct the annual-cycle model and then linearly regress the time-series against it. This would, for example, let us obtain statistical significance (p-value) and confidence intervals.

To construct the model, we directly extract coefficients from the sine and cosine regression outputs to enter the formula in Equation 2.12, as follows.

```
ann.cyc <- coef(lm.cos)[2]*cos(2*pi*(1/12)*tb) +
coef(lm.sin)[2]*sin(2*pi*(1/12)*tb) + mean(nottem)
```

We can regress the time-series against this reference sinusoidal model, as follows. The first command performs the linear regression and the second and third print the summary and confidence information (default 95%) respectively.

```
lm.ac <- lm(nottem ~ ann.cyc)
summary(lm.ac)
confint(lm.ac)
```

The intercept is 0 (*i.e.* the origin) and the gradient is 1, because we constructed a sinusoid of the correct amplitude and phase around the mean value. The confidence interval around the gradient is [0.9605, 1.0395]. If we want the confidence interval around the actual amplitude, we need to multiply these lower and upper confidence limits by the amplitude, as follows. The first command calculates the amplitude, the second prints the scaled confidence interval.

```
amp.ac <- sqrt(coef(lm.cos)[2]^2 + coef(lm.sin)[2]^2)
confint(lm.ac)[2,]*amp.ac
```

This reveals a 95% confidence interval of [11.10, 12.01] around the amplitude, *i.e.* at 95% confidence we have $11.10 \le |A_{20}| \le 12.01$ with a central estimate of $|A_{20}| = 11.56$.

If we were to repeat this procedure for other frequencies, we would find that the next most important frequency is f_{40} , *i.e.* the second harmonic of the annual sinusoid.

2.6 What Correlation and Regression 'See'

Figure 2.13 shows what a correlation of nottem against ann.cyc 'sees', the data are shown as the grey line and the modelled annual sinusoidal cycle as

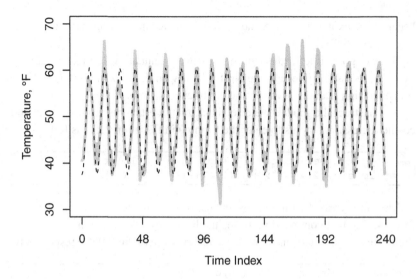

Figure 2.13 The nottem – annual sinusoid correlation.

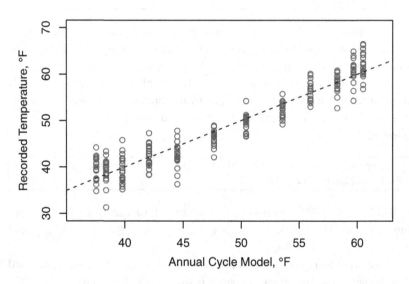

Figure 2.14 The nottem – annual cycle regression.

the dashed line. In Figure 2.13, the consistency of the same-sense, in-step, proportional behaviour is clear.

Figure 2.14 shows what a simple linear regression of nottem against ann.cyc 'sees', the recorded data are plotted against the cyclic modelled values in grey points (circles), with the regression line shown dashed. Note that the regression line has gradient equal to 1 and (extended to the left, not shown) passes through the origin, as revealed by the summary regression output. Also, and as indicated in the summary regression output, the differences between the points and the regression line (*i.e.* the residuals) are reasonably symmetrically distributed around the regression line, which is an indicator that the residuals are reasonably normally distributed. If necessary, we could examine the distribution of the residuals in more detail, *e.g.* by using standard statistical techniques to investigate the normality of their distribution.

2.7 Summary

We have shown that we can use cross-correlation to investigate pairs of time-series for similar features which occur in both time-series but shifted in time (lagged) with respect to each other. We have also shown that we can use autocorrelation to investigate a time-series for recurrent and periodic features.

We have developed this basic analysis to show that we can use correlation and the related technique of simple linear regression to compare time-series to 'reference' time-series, in particular reference waveforms comprising sinusoids at frequencies of $k = 1, 2, 3, 4 \ldots$ cycles over the duration, T, of the time-series, and obtain the proportion of variance explained (*i.e.* effect-size, coefficient of determination, R^2 or as the square of the correlation coefficient, R) and, additionally with linear regression, the amplitude of a sinusoidal component.

This is a solid basis from which to proceed with a consideration of the DFT but we first need to consider Fourier series in Chapter 3 and a little of the theory of the (analytical or continuous-time) Fourier transform in the first half of Chapter 4.

Exercises

1. Autocorrelate the beryllium time-series, using various values of the lag.max parameter. To what extent do the autocorrelograms confirm the periods revealed in sunspot.year autocorrelogram in Figure 2.9?
2. Autocorrelate the sunspot.month time-series, using various values of the lag.max parameter. To what extent do the autocorrelograms confirm the periods revealed in sunspot.year autocorrelogram in Figure 2.9?

3. There are two environmental time-series included with R, lynx in the
   datasets package and hare in the TSA library package that, despite being
   for different (neighbouring) regions of Canada, potentially reveal evidence
   of a predator-prey relationship. In summary, the lynx time-series is annual
   numbers of lynx trappings for 1821–1934 for the Mackenzie River region
   in Canada, the hare time-series is annual numbers of hare trappings for
   1905–1935 for the Hudson Bay catchment in Canada.

   (i) Autocorrelate both time-series and confirm the presence of *ca.*
       9.5-year cycles in the numbers of lynx and hare trapped annually.
   (ii) Extract sections from both time-series for the common period
        1905–1934 and cross-correlate to confirm a lag of 2 years for the
        variations in the lynx numbers behind the hare numbers.

4. Repeat the analysis of the example in Section 2.5 and confirm the presence
   of a second harmonic with frequency 2 cycles/year in the nottem dataset,
   i.e. $f_{40} = {}^{40}/_{240} = {}^{1}/_{6}$, having $R^2 = 0.0154$ (1.54%).

3

Fourier Series

In Chapter 2 we established that we could investigate a time-series for periodic (and other) features by generating a set of reference time-series of the same length (duration) and then systematically cross-correlating or linearly regressing the time-series under investigation against those reference time-series. We also observed that this would be a time-consuming procedure to follow in practice.

Fortunately, an alternative technique exists, namely the discrete Fourier transform (DFT) and, although this only applies for equal-interval time-series and specified sets of sinusoidal reference time-series, the DFT allows us to investigate periodic features in such time-series very rapidly. However, we first need to develop some more general Fourier theory, starting with Fourier series, to demonstrate and clarify what the DFT can – and cannot – do for us in the context of time-series analysis.

Starting with Fourier series is also useful because it builds from the correlation approaches in the previous chapter. In essence, the formulae for extracting the Fourier series coefficients, as in the following section, are correlations of the time-function with sines and cosines. The DFT is not explicitly defined in terms of correlations and is of use in many more contexts than identifying periodic features in time-series. However, in the context of identifying periodic features in time-series, it might be helpful to keep in mind that the DFT can be considered as being a short-cut that saves us from having to explicitly generate a particular set of reference sinusoids and then cross-correlating or linearly regressing our time-series of interest against every sinusoid in that reference set.

3.1 What Are Fourier Series?

A Fourier series is a representation of a continuous function as a sum of sinusoids. More formally, a Fourier series is a decomposition of a (periodic) continuous function as a linear combination of sinusoids, *i.e.* a set of sinusoids which when added together exactly reproduce the original function, including a zero-frequency term corresponding to the mean value of the data. All the non-zero-frequency sinusoids in the set have frequencies which are integer multiples of the fundamental frequency: the fundamental frequency is one cycle over the period of the function. Thus, there is a constant frequency interval, equal to the fundamental frequency, between adjacent frequencies in a Fourier series.

There are various equivalent ways of expressing the Fourier series formulae but the form in Equation 3.1 is reasonably standard in terms of general (angular) variable, θ, for function $x(\theta)$ having period 2π radians, *i.e.*

$$x(\theta) = \frac{a_0}{2} + \sum_{k=1}^{\infty} a_k \cos(k\theta) + \sum_{k=1}^{\infty} b_k \sin(k\theta) \tag{3.1}$$

for integers $k = 1, 2, 3, \ldots$ and $a_0/2$ is a zero-frequency term representing the mean level. Thus, k is the frequency in terms of an integer multiple (harmonic) of the fundamental frequency of one cycle over period 2π.

The coefficients a_k, b_k are given by the following formulae:

$$a_k = \frac{1}{\pi} \int_0^{2\pi} x(\theta) \cos(k\theta)\, d\theta$$

$$b_k = \frac{1}{\pi} \int_0^{2\pi} x(\theta) \sin(k\theta)\, d\theta$$

and

$$a_0 = \frac{1}{\pi} \int_0^{2\pi} x(\theta) \cos(0)\, d\theta = \frac{1}{\pi} \int_0^{2\pi} x(\theta)\, d\theta \tag{3.2}$$

which is consistent with a_k but evaluates to twice the mean value of $x(\theta)$ over the period, hence the division by 2 in Equation 3.1.

3.2 Orthogonality

The formulae for the coefficients in Equation 3.2 work because of the orthogonality conditions for sines and cosines on the interval $[0, 2\pi]$. To clarify, for all integers $n, m \neq 0$ we have the following orthogonality conditions:

$$\int_0^{2\pi} \cos(n\theta) \cos(m\theta) \, d\theta = \begin{cases} \pi, & n = m \\ 0, & n \neq m \end{cases}$$

$$\int_0^{2\pi} \sin(n\theta) \sin(m\theta) \, d\theta = \begin{cases} \pi, & n = m \\ 0, & n \neq m \end{cases}$$

$$\int_0^{2\pi} \sin(n\theta) \cos(m\theta) \, d\theta = 0, \qquad \forall n, m. \tag{3.3}$$

3.2.1 Clarification of Orthogonality

In familiar geometrical terms, two (straight) lines are orthogonal if they lie at right-angles to each other in the plane. Extending this, if we plot two variables with respect to some general, *e.g.* Cartesian, coordinate axes and they each yield a (straight) line and those lines are at right angles to each other, then the variables are orthogonal. In more formal terms, two variables are orthogonal if their inner (dot) product is zero: Equations 3.3 are examples of inner products.

Orthogonal variables are linearly independent, *i.e.* uncorrelated. In statistical terms, if we correlate a pair of orthogonal variables then we would obtain correlation coefficient $R = 0$ and, similarly, if we linearly regress a variable against another variable orthogonal to it then we would obtain coefficient of determination $R^2 = 0$. Conversely, non-orthogonal variables are linearly dependent, *i.e.* correlated. In statistical terms, if we correlate or linearly regress a pair of non-orthogonal variables then we would obtain $R \neq 0$ or $R^2 \neq 0$ respectively. We can extend the concept of orthogonality to a general set of N variables: if every variable in the set is orthogonal to every other variable in the set, then the set is termed orthogonal.

With regard to Fourier series, we have an orthogonal set of sinusoids of the form $a_k \cos(k\theta) + b_k \sin(k\theta)$, for integer k: any sinusoid in this set (for a given value of k) is orthogonal to any other sinusoid in the set (for any other value of k). In statistical terms, the sinusoids in the set are linearly independent and if we were to correlate or linearly regress any two of them (with different integer values of k) we would obtain $R = 0$ or $R^2 = 0$ respectively. This confirms our observations at the end of Chapter 2.

3.2.2 Alternative Justification of Fourier Series

We can demonstrate the reasonableness of Fourier series by starting with the conjecture that we *can* express the function $x(\theta)$ in Equation 3.1 as an explicit set of harmonic sinusoids, *i.e.*

$$x(\theta) = \sum_{k=0}^{\infty} (\alpha_k \cos(k\theta) + \beta_k \sin(k\theta)) \tag{3.4}$$

where α_k, β_k are general coefficients and k is an integer as specified above for the Fourier series (*i.e.* α_0 represents the zero-frequency mean level in this expression, corresponding to $a_0/2$ in Equation 3.1).

From that conjecture and noting the orthogonality conditions, it is clear that for any integer k, if we multiply the function by $\cos(k\theta)$ or $\sin(k\theta)$ and integrate over the interval $[0, 2\pi]$ then we will only obtain non-zero results for the $\cos(k\theta)$ or $\sin(k\theta)$ sinusoids, respectively, inside the summation. In other words, we can explicitly select terms according to their frequency (for k as specified).

From there, it is straightforward to deduce that it does not matter whether we consider the function $x(\theta)$ as an implicit set of harmonic sinusoids, with frequencies given by k as in Equation 3.1, or as an explicit set as in Equation 3.4: the same frequency selection must apply in both considerations. However, note that owing to the orthogonality criteria of the Fourier series, this only applies for this specific set of harmonic frequencies.

3.3 Fourier Series for General Time Function

For a general time function $x(t)$, we have to equate T, the duration of $x(t)$, to the period of the sinusoid, 2π, with exactly one cycle over that duration, *i.e.* with reference to Equation 3.1, we have $T = 2\pi$. Hence, the fundamental (first harmonic) frequency is $f_1 = 1/T$ and so the constant frequency interval is $\Delta f = f_1 = 1/T$. All the other non-zero-frequencies in a Fourier series are harmonics (integer multiples) of the fundamental frequency, *i.e.* the second harmonic is $f_2 = f_1 + \Delta f = 2f_1$, third harmonic is $f_3 = f_2 + \Delta f = 3f_1$ and so on with general kth harmonic $f_k = f_{k-1} + \Delta f = kf_1$. This enables us to write the general form for the Fourier series of a time function $x(t)$ of period T:

$$\begin{aligned}
x(t) &= \frac{a_0}{2} + \sum_{k=1}^{\infty} a_k \cos\left(\frac{2\pi k}{T}t\right) + \sum_{k=1}^{\infty} b_k \sin\left(\frac{2\pi k}{T}t\right) \\
&= \frac{a_0}{2} + \sum_{k=1}^{\infty} a_k \cos(2\pi f_k t) + \sum_{k=1}^{\infty} b_k \sin(2\pi f_k t)
\end{aligned} \tag{3.5}$$

for integers $k = 1, 2, 3, \ldots$, $f_k = k/T = kf_1$. Correspondingly, the coefficients a_k, b_k are given by the following formulae:

$$a_k = \frac{2}{T} \int_0^T x(t) \cos\left(\frac{2\pi k}{T}t\right) dt = \frac{2}{T} \int_0^T x(t) \cos(2\pi f_k t) \, dt$$

$$b_k = \frac{2}{T} \int_0^T x(t) \sin\left(\frac{2\pi k}{T}t\right) dt = \frac{2}{T} \int_0^T x(t) \sin(2\pi f_k t) \, dt$$

$$a_0 = \frac{2}{T} \int_0^T x(t) \, dt. \tag{3.6}$$

Note that the Fourier series sums are infinite, *i.e.* there is no finite maximum frequency.

Also, the function being decomposed must be integrable on the interval $[0, T]$, *i.e.* the integral

$$\int_0^T x(t) \, dt$$

must exist (must be finite). In practice, this means that functions must be either continuous or, if discontinuous, have a finite number of finite discontinuities and be finite everywhere (*i.e.* piecewise continuous).

3.3.1 Alternative Justification Revisited

If we return to the illustration in the preceding section but in terms of time, we can represent $x(t)$ as a sum of general sinusoids having frequencies which are integer multiples of the fundamental frequency and with arbitrary phase, *i.e.*

$$x(t) = \sum_{k=0}^{\infty} A_k \cos(2\pi f_k t - \phi_k) \tag{3.7}$$

where A_k, f_k and ϕ_k are respectively amplitude, frequency and phase constants for integers $k = 0, 1, 2, \ldots$ (*i.e.* A_0 is the mean level). From there, we can expand Equation 3.7 as we did Equation 2.11 to obtain

$$x(t) = A_0 + \sum_{k=1}^{\infty} A_k \cos(\phi_k) \cos(2\pi f_k t)$$

$$+ \sum_{k=1}^{\infty} A_k \sin(\phi_k) \sin(2\pi f_k t). \tag{3.8}$$

Equating to coefficients in the Fourier series, we have

$$a_k = A_k \cos(\phi_k)$$

$$b_k = A_k \sin(\phi_k)$$

$$a_0 = 2A_0. \tag{3.9}$$

These results are consistent with those of Section 2.5.

3.3.2 Angular Frequency

In some contexts, Fourier series (and Fourier transforms) are expressed in terms of angular frequency, $\omega = 2\pi f$, e.g.

$$x(t) = \frac{a_0}{2} + \sum_{k=1}^{\infty} a_k \cos(\omega_k t) + \sum_{k=1}^{\infty} b_k \sin(\omega_k t) \qquad (3.10)$$

with coefficients

$$a_k = \frac{2}{T} \int_0^T x(t) \cos(\omega_k t)\, dt$$

$$b_k = \frac{2}{T} \int_0^T x(t) \sin(\omega_k t)\, dt$$

$$a_0 = \frac{2}{T} \int_0^T x(t)\, dt. \qquad (3.11)$$

However, for time-series analysis we are primarily interested in frequency (cycles per unit time) rather than angular frequency (radians per unit time), so this alternative representation is not of direct use to us for the most part. Also, working in angular frequencies does not offer any advantage when interpreting the DFT.

3.4 Symmetry

Preservation of symmetry is an important property of Fourier series: the symmetry of a Fourier series corresponds exactly to the symmetry of the decomposed function. A function $x(t)$ is:

- odd-symmetric if $x(t) = -x(-t)$, sometimes referred to as anti-symmetric;
- even-symmetric if $x(t) = x(-t)$.

In general, functions have mixtures of odd- and even-symmetry and any function can be written in terms of odd- and even-symmetric constituent functions. Consider, for the general function $x(t)$:

- an odd-symmetric constituent function, $x_o(t) = \frac{1}{2}(x(t) - x(-t))$;
- an even-symmetric constituent function, $x_e(t) = \frac{1}{2}(x(t) + x(-t))$;
- and, clearly, $x(t) = x_o(t) + x_e(t)$.

Sines and cosines are odd- and even-symmetric functions respectively, and so:

- an odd-symmetric function decomposes to a Fourier series comprised entirely of sines;

- an even-symmetric function decomposes to a Fourier series comprised entirely of cosines;
- a general function decomposes to a Fourier series comprised of sines and cosines, the sine and cosine terms corresponding to the odd- and even-symmetric parts of the function respectively.

These observations might be a little surprising at first glance. However, if we have an odd-symmetric function on the left-hand side (*e.g.* of Equation 3.4) then the series on the right-hand side must also be odd-symmetric and any cosine terms in the right-hand side would add some even-symmetry which would be inconsistent with the left-hand side. Similarly, an even function on the left-hand side cannot have odd-symmetric terms in the series on the right-hand side.

These basic symmetry properties carry forward into the Fourier transform and DFT.

3.4.1 Pure Odd- and Even-Symmetry

In order to illustrate Fourier series and preservation of symmetry, let us consider an illustrative basic function such as the square wave. In its simplest forms, a square wave can be either odd-symmetric, $s_o(t)$, or even-symmetric, $s_e(t)$, as shown in Figures 3.1 and 3.2 respectively. Note that a square wave over a finite time interval is piecewise continuous, *i.e.* has a finite number of finite discontinuities (two discontinuities per cycle), and so we can integrate to obtain the Fourier series coefficients.

If we work through the calculations to extract the coefficients, we obtain the following Fourier series for the odd- and even-symmetric unit-amplitude square waves $s_o(t)$ and $s_e(t)$:

$$s_o(t) = \frac{4}{\pi} \left(\sin(2\pi f_1 t) + \frac{1}{3} \sin(2\pi f_3 t) + \frac{1}{5} \sin(2\pi f_5 t) \right.$$
$$\left. + \frac{1}{7} \sin(2\pi f_7 t) + \cdots \right)$$

$$s_e(t) = \frac{4}{\pi} \left(\cos(2\pi f_1 t) - \frac{1}{3} \cos(2\pi f_3 t) + \frac{1}{5} \cos(2\pi f_5 t) \right.$$
$$\left. - \frac{1}{7} \cos(2\pi f_7 t) + \cdots \right). \tag{3.12}$$

where $f_1 = 1/T, f_3 = 3f_1, f_5 = 5f_1, f_7 = 7f_1, \ldots$

Note that the even harmonic coefficients are all zero. Figures 3.1 and 3.2 show the Fourier series up to the fourth non-zero term, *i.e.* fourth partial-sum (sum of the first four terms, $f_1 + f_3 + f_5 + f_7$), for $s_o(t)$ and $s_e(t)$ respectively.

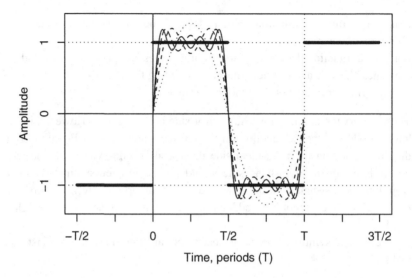

Figure 3.1 Odd-symmetry square wave, $s_o(t)$, with the Fourier series to the first four non-zero terms.

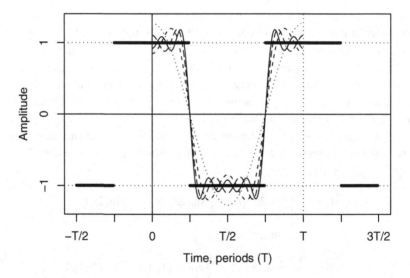

Figure 3.2 Even-symmetry square wave, $s_e(t)$, with the Fourier series to the first four non-zero terms.

In Figures 3.1 and 3.2, and Figure 3.3 in the following section, the heavy solid line is the square wave and the dotted, dashed, long-dashed and light solid lines are the first, second, third and fourth partial-sums respectively. It is clear in both odd- and even-symmetric cases that the Fourier series tends toward the square wave as the number of terms increases and the Fourier series will continue to further converge to the square wave as successive higher-order terms are included in the partial-sum. These two Fourier series show that:

- a pure odd-symmetric function decomposes to a series comprised entirely of sines, which are odd-symmetric functions;
- a pure even-symmetric function decomposes a series comprised entirely of cosines, which are even-symmetric functions.

In addition, these two examples show that both series contain only odd harmonics and the relative amplitudes of those harmonics are equal for square waves of the same amplitude and fundamental frequency, *i.e.* relative amplitudes 1, $1/3$, $1/5$, $1/7$, ... for the $f_1, f_3, f_5, f_7, \ldots$ terms in both cases. This hints at another important property of Fourier series, *i.e.*

- if we have the same function in different forms on the left-hand side which are only different with regard to their time (phase) reference, as we do with these odd and even square waves, then we have the same harmonic content across all the different forms of the function.

3.4.2 General Mixed Symmetry

We can explore these observations a little further and draw some general inferences by considering an arbitrary-phase square wave that is neither pure odd-symmetric nor pure even-symmetric. Calculating the coefficients is laborious for an arbitrary-phase square wave, and is rarely done for this reason (it is easier to adjust the phase to either odd- or even-symmetry), but is relatively tractable for the particular square wave, $s_h(t)$ as shown in Figure 3.3, which is exactly half-way between the odd- and even-symmetric square waves in the preceding section.

The Fourier series for this square wave, $s_h(t)$, is

$$s_h(t) = \frac{1}{\sqrt{2}} \cdot \frac{4}{\pi} \left(\sin(2\pi f_1 t) - \frac{1}{3} \sin(2\pi f_3 t) - \frac{1}{5} \sin(2\pi f_5 t) \right.$$

$$\left. + \frac{1}{7} \sin(2\pi f_7 t) + \cdots \right)$$

$$+ \frac{1}{\sqrt{2}} \cdot \frac{4}{\pi} \left(\cos(2\pi f_1 t) + \frac{1}{3} \cos(2\pi f_3 t) - \frac{1}{5} \cos(2\pi f_5 t) \right.$$

$$\left. - \frac{1}{7} \cos(2\pi f_7 t) + \ldots \right). \tag{3.13}$$

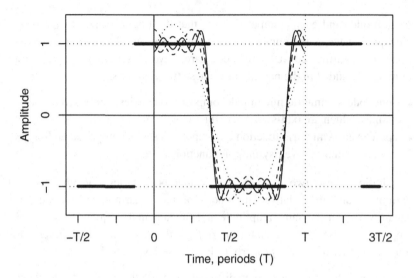

Figure 3.3 Mixed-symmetry square wave, $s_h(t)$, with the Fourier series to the first four non-zero terms.

As written in Equation 3.13, this Fourier series comprises two distinct parts, one of sine terms and one of cosine terms. These terms have frequencies in common and the series can be rearranged according to frequency as

$$
\begin{aligned}
s_h(t) = \frac{1}{\sqrt{2}} \cdot \frac{4}{\pi} \Big(& \big(\cos(2\pi f_1 t) + \sin(2\pi f_1 t) \big) \\
& + \frac{1}{3} \big(\cos(2\pi f_3 t) - \sin(2\pi f_3 t) \big) - \frac{1}{5} \big(\cos(2\pi f_5 t) + \sin(2\pi f_5 t) \big) \\
& - \frac{1}{7} \big(\cos(2\pi f_7 t) - \sin(2\pi f_7 t) \big) + \cdots \Big).
\end{aligned} \tag{3.14}
$$

This shows the same sequence of relative amplitudes *i.e.* 1, 1/3, 1/5, 1/7, ... for the first four non-zero terms in both forms. We can further simplify Equation 3.14 to either sine or cosine form to express the harmonic content as explicitly as possible and, for example, in sine form we have

$$
\begin{aligned}
s_h(t) = \frac{4}{\pi} \Big(& \sin\left(2\pi f_1 t + \frac{\pi}{4}\right) - \frac{1}{3} \sin\left(2\pi f_3 t - \frac{\pi}{4}\right) \\
& - \frac{1}{5} \sin\left(2\pi f_5 t + \frac{\pi}{4}\right) + \frac{1}{7} \sin\left(2\pi f_7 t - \frac{\pi}{4}\right) + \cdots \Big).
\end{aligned} \tag{3.15}
$$

This is clearly a Fourier series of sinusoidal terms at only odd harmonics.

3.4.3 Symmetry and Phase

The three previous examples, which are the same square wave but with different phase (*i.e.* different symmetry relative to $t = 0$) taken together show that:

- a mixed odd-/even-symmetric function decomposes to a Fourier series comprised of sine and cosine terms, representing the odd- and even-symmetric parts of the function respectively;
- all three series contain the same harmonic structure irrespective of their different phases, *i.e.* the same frequencies with the same relative amplitudes (*i.e.* relative amplitude 1 at frequency f_1, $1/3$ at f_3, $1/5$ at f_5 and $1/7$ at f_7 for the first four terms).

3.5 General Properties of Fourier Series

3.5.1 Symmetry and Harmonic Composition

The preceding observations allow us to deduce the following general properties of Fourier series:

- a general function decomposes to a Fourier series comprised of sine and cosine terms;
- the sine and cosine terms represent the odd- and even-symmetric parts of the function respectively;
- the phase of the function does not affect the harmonic content, *i.e.* does not affect the amplitudes of the harmonics, but does affect the symmetry and so determines the balance between sine and cosine terms in the Fourier series.

3.5.2 Linearity

Also, as well as preserving symmetry and harmonic content, Fourier series are linear: *i.e.* the addition and scalar multiplication properties apply to Fourier series. If we have two general time functions $x(t)$ and $x'(t)$ with Fourier series, *i.e.*

$$x(t) = \frac{a_0}{2} + \sum_{k=1}^{\infty} a_k \cos(2\pi f_k t) + \sum_{k=1}^{\infty} b_k \sin(2\pi f_k t)$$

$$x'(t) = \frac{a'_0}{2} + \sum_{k=1}^{\infty} a'_k \cos(2\pi f_k t) + \sum_{k=1}^{\infty} b'_k \sin(2\pi f_k t) \qquad (3.16)$$

then:

- if we have $y(t) = Ax(t)$, where A is a constant, then the Fourier series for $y(t)$ is A times the Fourier series for $x(t)$, i.e.

$$y(t) = \frac{Aa_0}{2} + \sum_{k=1}^{\infty} Aa_k \cos(2\pi f_k t) + \sum_{k=1}^{\infty} Ab_k \sin(2\pi f_k t) \quad (3.17)$$

- if we have $y(t) = x(t) + x'(t)$ then the Fourier series for $y(t)$ is the sum of the individual Fourier series for $x(t)$ and $x'(t)$, i.e.

$$y(t) = \frac{(a_0 + a_0')}{2} + \sum_{k=1}^{\infty} (a_k + a_k') \cos(2\pi f_k t)$$

$$+ \sum_{k=1}^{\infty} (b_k + b_k') \sin(2\pi f_k t) \quad (3.18)$$

These are important properties of Fourier series which carry forward into the DFT.

3.6 Complex Fourier Series

In order to demonstrate the derivation of the Fourier transform and DFT, or at least to do this with any degree of convenience, we must move from a (real) Fourier series of paired sine and cosine terms to a complex Fourier series of complex exponential terms. To do this, we make use of Euler's formulae, i.e.

$$e^{i\theta} = \cos\theta + i\sin\theta$$
$$e^{-i\theta} = \cos\theta - i\sin\theta. \quad (3.19)$$

We can rearrange these for sine and cosine explicitly, i.e.

$$\cos\theta = \frac{1}{2}\left(e^{i\theta} + e^{-i\theta}\right)$$

$$\sin\theta = \frac{-i}{2}\left(e^{i\theta} - e^{-i\theta}\right). \quad (3.20)$$

From there, we can substitute for $\cos(2\pi f_k t)$ and $\sin(2\pi f_k t)$ terms in the Fourier series formula and rearrange as follows:

$$x(t) = \frac{a_0}{2} + \sum_{k=1}^{\infty} \left(\frac{a_k}{2} \left(e^{2\pi i f_k t} + e^{-2\pi i f_k t} \right) - \frac{ib_k}{2} \left(e^{2\pi i f_k t} - e^{-2\pi i f_k t} \right) \right)$$

$$= \frac{a_0}{2} + \sum_{k=1}^{\infty} \frac{1}{2}(a_k - ib_k) e^{2\pi i f_k t} + \sum_{k=1}^{\infty} \frac{1}{2}(a_k + ib_k) e^{-2\pi i f_k t}$$

$$= \frac{a_0}{2} + \sum_{k=1}^{\infty} \frac{1}{2} (a_k - ib_k) e^{2\pi i f_k t} + \sum_{k=-1}^{-\infty} \frac{1}{2} (a_{-k} - ib_{-k}) e^{2\pi i f_k t}$$

$$= \sum_{k=-\infty}^{\infty} c_k e^{2\pi i f_k t}. \tag{3.21}$$

The complex coefficients relate directly to the real coefficients, and noting positive–negative complex-conjugate frequency symmetries ($a_{-k} = a_k$, $b_{-k} = -b_k$ and $b_0 = 0$), we have:

$$c_k = \frac{1}{2} (a_k - ib_k)$$

$$= \frac{1}{T} \int_0^T x(t) \left(\cos (2\pi f_k t) - i \sin (2\pi f_k t) \right) dt$$

$$= \frac{1}{T} \int_0^T x(t) e^{-2\pi i f_k t} dt. \tag{3.22}$$

3.6.1 Negative Frequencies

It is evident that the complex Fourier series contains negative frequencies, arising from the representation of real sines and cosines in terms of complex exponentials. This can be initially perplexing, but remember there is a crucial difference between the real and complex Fourier series representations:

- real Fourier series – representation in terms of real sinusoids;
- complex Fourier series – representation in terms of complex exponentials;

and, although they are inter-convertible, real sinusoids and complex exponentials are not the same thing.

In this context, it is fine to consider negative frequencies as a mathematical convenience which streamlines the notation and algebraic manipulation. Also, do not forget that if we back-substitute from complex exponentials to real sinusoids, we obtain a representation entirely in terms of positive frequencies again.

3.6.2 Real Data and Frequency Components

From Equations 3.21 and 3.22, we can expand the general kth positive and negative frequency coefficients to obtain explicit sinusoidal forms for the positive and negative kth terms in the complex Fourier series (positive and negative kth frequency components):

- general positive *kth* frequency component, f_k, coefficient $c_k = \frac{1}{2}(a_k - ib_k)$ has form

$$x_k(t) = \frac{a_k}{2}\cos(2\pi f_k t) - i\frac{b_k}{2}\sin(2\pi f_k t) \qquad (3.23)$$

- general negative *kth* frequency component, $-f_k = f_{-k}$, coefficient $c_{-k} = \frac{1}{2}(a_{-k} - ib_{-k})$ has form

$$
\begin{aligned}
x_{-k}(t) &= \frac{-a_k}{2}\cos(2\pi f_{-k}t) - i\frac{-b_k}{2}\sin(2\pi f_{-k}t) \\
&= \frac{a_k}{2}\cos(2\pi f_k t) + i\frac{b_k}{2}(-\sin(2\pi f_k t)) \\
&= \frac{a_k}{2}\cos(2\pi f_k t) - i\frac{b_k}{2}\sin(2\pi f_k t) \\
&= x_k(t) \qquad\qquad\qquad\qquad\qquad\qquad\qquad (3.24)
\end{aligned}
$$

Thus, the complex Fourier series for real data divides the effect-size (amplitude, energy, power) of the *kth* frequency component in the time-function equally across the positive and negative *kth* frequency terms in the complex Fourier series, as expected from the form of the complex Fourier coefficients. This positive–negative frequency symmetry for real data carries through to the DFT.

3.6.3 Complex Data

Consideration of complex data is beyond the scope of this book but, in principle, if we were to perform a real Fourier series decomposition of complex data then we would obtain separate real Fourier series for the real and imaginary parts. We could then combine these two series term-wise according to frequency into a single series having real and imaginary parts for each frequency term.

Also, if we were to perform a complex Fourier series decomposition of complex data then we would obtain separate series of complex exponential terms for the real and imaginary parts. We could then combine these two series term-wise according to frequency into a single series having real and imaginary parts for each frequency term.

In terms of the symmetries of positive and negative frequency components, for complex data the situation is similar to that for real data but more complicated as we also have to consider the conjugate-symmetries of the imaginary part. These can be considered as a mirror-image of the conjugate-symmetries for the real part. However, we would still be able (at least in principle) to obtain all the real sinusoidal coefficients from the complex coefficients and back-substitute from the series of complex exponentials to obtain real Fourier series for the real and imaginary parts.

3.7 Summary

A real Fourier series represents a continuous or piecewise continuous function as a sum of sinusoids. The sinusoids have discrete frequencies which are harmonics – integer multiples – of the fundamental frequency which is one cycle over the period of the function. The frequency difference between adjacent terms in the series is constant and equal to the fundamental frequency. The phase of the function does not affect the harmonic content but does affect the symmetry and so controls the balance between sine and cosine terms in the Fourier series.

Alternatively, the real Fourier series can be considered as an infinite spectrum of discrete frequencies, all of which are harmonics – integer multiples – of the fundamental frequency.

The complex Fourier series is an alternative representation of the real Fourier series in terms of complex exponentials. The complex representation simplifies the extension of the Fourier series to the Fourier transform and DFT.

Exercises

1. Derive the Fourier series coefficients for the odd-symmetric square wave.
2. Derive the Fourier series coefficients for the even-symmetric square wave.
3. Verify the orthogonality conditions by directly evaluating the integrals in Section 3.2.

4

Fourier Transforms

In the previous chapter we established that we could use Fourier series to represent a continuous time function as a linear combination of harmonic sinusoids. In other words, we have considered a Fourier series as being a decomposition of a time function of period T in terms of a sum of sinusoidal components which have frequencies that are harmonics of the fundamental frequency, $f_1 = 1/T$. However, although the Fourier series coefficients themselves are explicitly frequency terms, we have not considered the set of coefficients in terms of being a frequency function (*i.e.* a 'spectrum').

4.1 Frequency Functions

If we wished, we could construct an explicit frequency function, $X(f)$, from the Fourier series coefficients for a time function, $x(t)$, and this would be a discrete frequency-domain representation of $x(t)$. We would then have a relationship between the time and frequency domains in the sense that we could completely represent any time function as a frequency function by making use of its harmonic composition.

Starting with the time function, *i.e.* real Fourier series as in Equations 3.5, we have

$$x(t) = \frac{a_0}{2} + \sum_{k=1}^{\infty} (a_k \cos(2\pi f_k t) + b_k \sin(2\pi f_k t)) \tag{4.1}$$

or, for complex Fourier series, Equation 3.20, we have

$$x(t) = \sum_{k=-\infty}^{\infty} c_k e^{2\pi i f_k t}. \tag{4.2}$$

46

We know both the frequencies for $k = 1, 2, 3, \ldots$ and the forms that the terms in the summation must take, which means that the frequency function can be simplified to the ordered (indexed) list of coefficients, *i.e.* $\{(a_k, b_k)\}$ or $\{c_k\}$ for real and complex Fourier series respectively. More completely, the frequency function is, in terms of the real Fourier series coefficients

$$X(f) = \frac{a_0}{2} + \{(a_k, b_k)\}_{k=1}^{\infty} \qquad (4.3)$$

or, in terms of the complex Fourier series coefficients

$$X(f) = \{c_k\}_{k=-\infty}^{\infty} . \qquad (4.4)$$

If we know all the coefficients then we can completely reconstruct the original function, $x(t)$, and so $X(f)$ is a frequency-domain representation of $x(t)$ according to the Fourier series formalisation.

Both forms in Equations 4.3 and 4.4 are discrete frequency functions because the finite duration of the time function, T, restricts the lowest non-zero frequency we can resolve to $f_1 = 1/T$ and higher frequencies are restricted to integer multiples (harmonics) of this fundamental frequency because of the orthogonality conditions (Equations 3.4).

We will start from here in our derivation of the discrete Fourier transform (DFT) but, first, we will briefly consider the (analytical or continuous-time) Fourier transform to highlight some important general features of Fourier spectra.

4.2 The Fourier Transform

The Fourier transform is the limiting case of the complex Fourier series where:

- the period tends to infinity, *i.e.* $T \to \infty$, and thus
- the finite frequency interval, $\Delta f = 1/T$, tends to the infinitesimally small, *i.e.* $\Delta f \to df$.

This requires that the Fourier series summations for finite Δf become Fourier integrals for infinitesimal df. Stated without formal derivation, this yields a pair of infinite integrals: the forward transform, from time to frequency ($-i$ in the complex exponential), and the inverse transform, from frequency to time ($+i$ in the complex exponential), *i.e.*

$$X(f) = \int_{-\infty}^{\infty} x(t)\, e^{-2\pi i f t} dt$$

$$x(t) = \int_{-\infty}^{\infty} X(f)\, e^{2\pi i f t} df \qquad (4.5)$$

respectively.

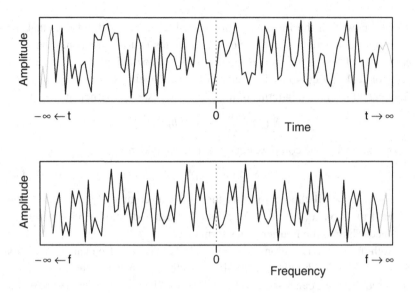

Figure 4.1 Time (upper) and frequency (lower) functions.

The frequency function, $X(f)$, is continuous. It can be considered as a continuous spectrum of frequencies in which the discrete frequency coefficients in a Fourier series have merged into a continuous frequency function as the frequency interval between adjacent frequencies tends to zero (and the number of frequency coefficients in any given finite frequency interval, however small, tends to infinity). This is shown schematically in Figure 4.1, which shows central regions of (a) an infinite real time function (in the upper plot) and (b) the amplitude of the infinite (complex) Fourier transform frequency function (in the lower plot).

Note the symmetry around zero in the frequency function in Figure 4.1: this positive–negative frequency symmetry is characteristic of the Fourier transforms of real functions and we will discuss this in more detail in Sections 4.4 and 4.7. Also note that although we are considering time–frequency relationships, the Fourier transform and Fourier series are not limited to time and frequency. If we are investigating periodic features in, for example, spatial data, we can transform between distance functions (with distance as a time-analogue in the formulae in Equation 4.5) and spatial-frequency (cycles per unit distance) functions.

4.2.1 Fourier Transform Notation

A commonly used notation, which saves us from explicitly writing the integrals where we just need to indicate that functions have a forward or inverse Fourier transform relationship, is to use a cursive capital $\mathcal{F}$ for the Fourier transform operator, *i.e.*

- the forward transform, from time to frequency, $X(f) = \mathcal{F}(x(t))$,
- the inverse transform, from frequency to time, $x(t) = \mathcal{F}^{-1}(X(f))$.

4.2.2 Angular Frequency

As with Fourier series, the Fourier transform pair is sometimes written in terms of the angular frequency, $\omega = 2\pi f$, in which case we need to include a normalising factor of $1/2\pi$ in the inverse transform due to the change of variable. This gives the forward and inverse integrals,

$$X(\omega) = \int_{-\infty}^{\infty} x(t) e^{-i\omega t} dt$$

$$x(t) = \frac{1}{2\pi} \int_{-\infty}^{\infty} X(\omega) e^{i\omega t} d\omega \tag{4.6}$$

respectively.

This form is advantageous in many circumstances but does not help with the derivation, use or interpretation of the DFT in the context of time-series analysis and so we will continue to use the frequency form.

4.2.3 Integrability

The Fourier integrals are infinite integrals and there are conditions on $x(t)$ and $X(f)$ such that the integrals can be evaluated. These are essentially the Dirichlet conditions, named after J P G L Dirichlet (1805–1859) who was instrumental in verifying and developing Fourier's analysis. Stated informally and without proof, although they are intuitively reasonable, these conditions are:

- $x(t) \to 0$ as $t \to \pm\infty$ and $X(f) \to 0$ as $f \to \pm\infty$ or, more formally
- $x(t)$ and $X(f)$ must be square-integrable, *i.e.* the integrals

$$\int_{-\infty}^{\infty} |x(t)|^2 dt \text{ and } \int_{-\infty}^{\infty} |X(f)|^2 df$$

must both exist (*i.e.* be finite), which implies that $x(t)$ and $X(f)$ must be (piecewise) continuous with respect to t and f respectively;
- $x(t)$ and $X(f)$ should be single-valued with respect to t and f respectively.

However, note that we cannot directly evaluate the Fourier transform of an infinite sinusoid because this does not satisfy these integrability criteria.

In order to Fourier transform sinusoids, we require a special type of function, *i.e.* the Dirac δ-function, which is introduced in Section 4.4.

4.3 General Properties of the Fourier Transform

The general properties of the Fourier transform correspond closely to the general properties of the Fourier series considered in the previous chapter – and also carry forward into the DFT.

4.3.1 Symmetry

The symmetry properties are:

- the Fourier transform of a real odd-symmetric function is pure imaginary, corresponding to the sine part of a (complex) Fourier series;
- the Fourier transform of a real even-symmetric function is pure real, corresponding to the cosine part of a (complex) Fourier series;
- the Fourier transform of a real mixed-symmetry function has real and imaginary parts, the real part corresponds to the even-symmetric part of the function, the imaginary part corresponds to the odd-symmetric part.

4.3.2 Linearity

As is the case with Fourier series, Fourier transforms are linear. If we have two general time functions $x(t)$ and $x'(t)$, then their Fourier transforms are

$$X(f) = \int_{-\infty}^{\infty} x(t) e^{-2\pi i f t} dt$$

$$X'(f) = \int_{-\infty}^{\infty} x'(t) e^{-2\pi i f t} dt. \tag{4.7}$$

Therefore, if we have $y(t) = Ax(t) + Bx'(t)$ where A and B are (scaling) constants, then the Fourier transform for $y(t)$ is the sum of the scaled individual Fourier transforms for $x(t)$ and $x'(t)$, *i.e.*

$$Y(f) = \int_{-\infty}^{\infty} y(t) e^{-2\pi i f t} dt = \int_{-\infty}^{\infty} \left(Ax(t) + Bx'(t)\right) e^{-2\pi i f t} dt$$

$$= A \int_{-\infty}^{\infty} x(t) e^{-2\pi i f t} dt + B \int_{-\infty}^{\infty} x'(t) e^{-2\pi i f t} dt$$

$$= AX(f) + BX'(f). \tag{4.8}$$

Or, more succinctly,

$$\mathcal{F}\left(Ax\,(t) + Bx'\,(t)\right) = A\mathcal{F}\,(x\,(t)) + B\mathcal{F}\left(x'\,(t)\right). \tag{4.9}$$

The inverse transforms are also linear.

4.4 Fourier Transforms of Sinusoids

A little consideration shows that the Fourier series for an arbitrary sinusoid of frequency k cycles over period T, e.g. $x\,(t) = A_k \cos\,(2\pi f_k t - \phi_k)$, is the sinusoid itself. If we evaluate the real Fourier series, we obtain a single pair of sine and cosine coefficients (a_k, b_k) for the sinusoid $a_k \cos\,(2\pi f_k t) + b_k \sin\,(2\pi f_k t)$, which rearranges to the original sinusoid. Similarly, if we evaluate the complex Fourier series, we obtain the single pair of complex coefficients (c_k, c_{-k}) for the complex exponential $c_k e^{2\pi i f_k t} + c_{-k} e^{-2\pi i f_k t}$ which rearranges to the original sinusoid.

Thus, it is reasonable to expect that the Fourier transform of an infinite sinusoid $x\,(t) = A_k \cos\,(2\pi f_k t - \phi_k)$ will be a single pair of frequency 'markers' at $\pm f_k$ ($f_{\pm k}$) corresponding to coefficients $c_{\pm k}$ in the complex Fourier series. However, if we try to evaluate the Fourier transform for this sinusoid we obtain

$$X\,(f) = \int_{-\infty}^{\infty} A_k \cos\,(2\pi f_k t - \phi_k)\, e^{-2\pi i f t} dt$$

$$= A_k \cos\,(\phi) \int_{-\infty}^{\infty} \cos\,(2\pi f_k t)\, e^{-2\pi i f t} dt$$

$$+ A_k \sin\,(\phi) \int_{-\infty}^{\infty} \sin\,(2\pi f_k t)\, e^{-2\pi i f t} dt. \tag{4.10}$$

As observed in Section 4.2.3, these integrals do not meet the integrability criteria, i.e. neither integral tends to zero (or any limit) as $t \to \pm\infty$, and so cannot be directly integrated to 'true' functions. Various mathematicians conjectured that the Fourier transform of a pure sinusoid, i.e. a function with single frequency f_k, must somehow be consistent with complex Fourier series and so contain a pair of markers at $\pm f_k$ and be zero everywhere else. However, it was only when P A M Dirac (1930) formalised the δ-function ('delta-function') that this situation was fully resolved.

4.4.1 The Dirac δ-Function

The Dirac δ-function is termed a *generalised function* which has the following basic properties (we are using frequency as the variable here, as frequency

is the variable required for the Fourier transform of a time-sinusoid, but
δ-functions can be in terms of any variable)

$$\delta(f) = \begin{cases} \infty, & f = 0 \\ 0, & f \neq 0 \end{cases} \text{ and } \delta(f - f_k) = \begin{cases} \infty, & f = f_k \\ 0, & f \neq f_k \end{cases}; \qquad (4.11)$$

$$\int_{-\infty}^{\infty} \delta(f)\, df = 1 \text{ and } \int_{-\infty}^{\infty} \delta(f - f_k)\, df = 1; \qquad (4.12)$$

$$\int_{-\infty}^{\infty} A\delta(f)\, df = A \text{ and } \int_{-\infty}^{\infty} \delta(Af)\, df = \frac{1}{|A|}; \qquad (4.13)$$

$$\int_{-\infty}^{\infty} X(f)\, \delta(f - f_k)\, df = X(f_k). \qquad (4.14)$$

The Dirac δ-function is to some extent counter-intuitive and requires a
little thought: it is an infinitesimally narrow spike of unit area, and so is
infinitely high. Also, a scaled δ-function *e.g.* $A\delta(f)$, is an infinitesimally
narrow infinitely high spike of area A. Thus, the scale-factor A is clearly
a measure of magnitude of a δ-function but care must be exercised when
considering it as an amplitude as would apply to a true function.

Perhaps the most intuitive way of considering the δ-function is as the
limiting case of the normal distribution as the width (standard deviation) tends
to zero, *i.e.*

$$\delta(f - f_k) = \lim_{\sigma \to 0} \frac{1}{\sigma\sqrt{2\pi}} e^{-\frac{(f - f_k)^2}{2\sigma^2}}, \qquad (4.15)$$

but this has limitations, *e.g.* the existence of the very small but non-zero
tails of the function as $f \to \pm\infty$ however small σ becomes. There are other
ways of considering the δ-function, *e.g.* as limiting cases of rectangles of unit
area as the width tends to zero but, whilst these resolve the tails issues, these
are not smooth continuous functions.

However, despite such counter-intuitive properties, the selection or sifting
property of the δ-function in Equation 4.14 allows us to develop a Fourier-
transform relationship between sinusoids and δ-functions which is consistent
with our expectation. If we consider Equation 4.14 with $X(f) = e^{2\pi i f t}$ then we
have the inverse Fourier transform of $\delta(f - f_k)$ and

$$x(t) = \int_{-\infty}^{\infty} \delta(f - f_k)\, e^{2\pi i f t} df = e^{2\pi i f_k t}. \qquad (4.16)$$

Thus, we have $e^{2\pi i f_k t} = \mathcal{F}^{-1}(\delta(f - f_k))$ and so we must have the following Fourier transform relationships,

$$\mathcal{F}\left(e^{2\pi i f_k t}\right) = \delta(f - f_k)$$

$$\mathcal{F}\left(e^{-2\pi i f_k t}\right) = \delta(f + f_k). \tag{4.17}$$

4.4.2 Fourier Transforms of Sines and Cosines

Having established the basic Fourier transform relationship between complex exponentials and δ-functions, we can use Euler's formula (Equation 2.17) to expand Equations 4.17, to obtain

$$\mathcal{F}(\cos(2\pi f_k t) + i\sin(2\pi f_k t)) = \delta(f - f_k)$$

$$\mathcal{F}(\cos(2\pi f_k t) - i\sin(2\pi f_k t)) = \delta(f + f_k). \tag{4.18}$$

Hence, rearranging and using the linearity properties in Equation 4.9, we obtain

$$\mathcal{F}(\cos(2\pi f_k t)) = \frac{1}{2}(\delta(f - f_k) + \delta(f + f_k))$$

$$\mathcal{F}(\sin(2\pi f_k t)) = \frac{i}{2}(\delta(f - f_k) - \delta(f + f_k)). \tag{4.19}$$

To summarise, therefore:

- the Fourier transform of a sine (frequency f_k, amplitude A) is a pair of odd-symmetric δ-functions (at $\pm f_k$, scale-factor $A/2$) in the imaginary part, as shown schematically in Figure 4.2;
- the Fourier transform of a cosine (frequency f_k, amplitude A) is a pair of even-symmetric δ-functions (at $\pm f_k$, scale-factor $A/2$) in the real part, as shown schematically in Figure 4.3.

4.4.3 Fourier Transform of a General Sinusoid

More generally, the Fourier transform of a general sinusoid of frequency f_k and amplitude A, e.g. $x(t) = A\cos(2\pi f_k t - \phi)$ as in Equation 2.9, comprises both a pair of odd-symmetric δ-functions at $\pm f_k$ with scale-factor $A_{Im} = \frac{A}{2}\sin(\phi)$ in the imaginary part, and a pair of even-symmetric δ-functions at $\pm f_k$ with scale-factor $A_{Re} = \frac{A}{2}\cos(\phi)$ in the real part, and $A = 2\sqrt{A_{Re}^2 + A_{Im}^2}$.

As an illustration of this general behaviour, the Fourier transform of the mixed-symmetry square wave $s_h(t)$, as considered in Section 3.4, is shown in

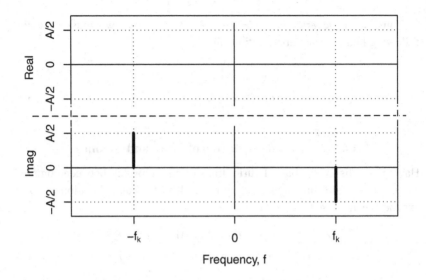

Figure 4.2 Fourier transform of $A \sin (2\pi f_k t)$.

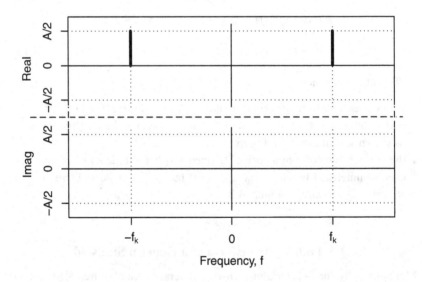

Figure 4.3 Fourier transform of $A \cos (2\pi f_k t)$.

Figure 4.4. This shows the relative amplitudes and phases of the first, third, fifth and seventh harmonics (note the overall scaling factor $4/\pi\sqrt{2} \approx 0.9$ in the amplitudes of the sinusoids in Equation 3.13.)

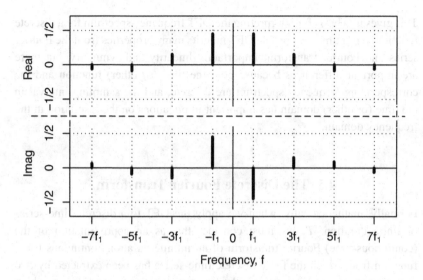

Figure 4.4 Fourier transform of $s_h(t)$.

4.4.4 From Continuous to Discrete Functions

There are many aspects of the Fourier transform that we have not covered, *e.g.* convolution and correlation, but which are covered in many books on Fourier theory. However, we have covered the main properties in the context of identifying periodic features in time-series (or other sequential data), which are properties that carry forward to the DFT.

The Fourier transform of a (piecewise) continuous function in the time domain is a frequency domain representation of it, *i.e.* a continuous function in the frequency domain. Fourier transforms are not restricted to time–frequency relationships: the Fourier transform of a (piecewise) continuous function in an arbitrary domain is a continuous function in the corresponding frequency domain. For example, the Fourier transform of a spatial function (with distance units such as metres) will be a spatial-frequency function (with frequencies in terms of cycles per unit distance, *e.g.* cycles per metre).

Alternatively, the Fourier transform of a function can be considered as an infinite continuous spectrum of frequencies contained in that function.

In terms of the identification of periodic features, if we have a time function of infinite duration comprised of sinusoids of frequencies $f_a, f_b, f_c, \ldots$ then the Fourier transform will yield a frequency spectrum comprised of complex pairs of δ-functions at $\pm f_a, \pm f_b, \pm f_c, \ldots$ ($f_{\pm a}, f_{\pm b}, f_{\pm c}, \ldots$), with magnitudes and signs corresponding to the amplitudes and phases of the sinusoids.

This gives us a basis for interpreting the DFT frequency spectrum for a discrete function, *e.g.* a time-series. The DFT inherits many properties from the Fourier series and Fourier transform, importantly linearity and symmetry, but there are important differences because both the time (or other) function and the corresponding frequency spectrum are discrete, and the sampling interval in the time (or other) domain has important implications on the resolution in the frequency domain.

4.5 The Discrete Fourier Transform

Formally, mathematically, when we apply the DFT to a discrete time-series of finite duration, T, we must consider this as an approximation of the (continuous-time) Fourier transform of an infinite-duration continuous time-function from which the finite discrete time-series has been extracted by two sampling processes. First, there is the 'top-hat' function (*i.e.* a function which is equal to 1 on interval $[0, T]$ and 0 everywhere else) which extracts the continuous sample of duration T from the infinite function. Second, there is the Dirac-comb (*i.e.* sequence of δ-functions Δt apart on interval $[0, T]$, where Δt corresponds to the time interval we considered in Section 2.2) which produces the discrete time-series of duration T from the continuous time-function of duration T.

This implied sampling process is shown diagrammatically in Figure 4.5.

Consequently, a rigorous, formal derivation of the DFT (of a discrete time-series) from the Fourier transform requires consideration of the convolutions of the Fourier transform (of a continuous time-function) with the Fourier transforms of the top-hat function and Dirac-comb. Whilst this full consideration is necessary for tasks such as signal processing, where the effects of the convolutions on the frequency spectrum are highly important, there is a more intuitive, albeit less formal, derivation from the complex Fourier series which is more informative in the context of time-series analysis.

4.5.1 Derivation of the DFT from Fourier Series

That more intuitive derivation of the DFT starts with the complex Fourier series frequency function in Equation 4.4 which, with coefficients re-expressed explicitly, is

$$X(f) = \left\{ \frac{1}{T} \int_0^T x(t)\, e^{-2\pi i f_k t} dt \right\}_{k=-\infty}^{\infty} = \{X_k\}_{k=-\infty}^{\infty} ; \qquad (4.20)$$

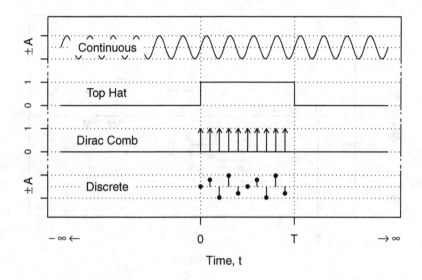

Figure 4.5 Continuous time-function to discrete time-series.

from which we consider a general kth harmonic term (frequency $f_k = k/T$)

$$X_k = \frac{1}{T} \int_0^T x(t)\, e^{-2\pi i f_k\, t} dt. \tag{4.21}$$

The time function, $x(t)$, in Equation 4.22 is continuous. If we sample $x(t)$ at a sequence of constant intervals starting at $t = 0$ we obtain a time-index as we considered in Section 2.2, *i.e.*

$$t = 0, \Delta t, 2\Delta t, \ldots, n\Delta t, \ldots, (N-1)\,\Delta t = \{t_n\}_{n=0}^{N-1} \tag{4.22}$$

where the constant sampling interval, $\Delta t = T/N$, is the reciprocal of the constant sampling frequency, $f_s = 1/\Delta t$, then we obtain a sequence of values $x(t_n)$ as a discrete approximation of $x(t)$, *i.e.*

$$x(t) \approx \{x(t_n)\}_{n=0}^{N-1} = \{x_n\}_{n=0}^{N-1} \tag{4.23}$$

for integers $n = 0, 1, \ldots, N-1$.

In the context of the detection of periodic features, we can consider the sampling process as effectively approximating the function $x(t)$ by a piecewise continuous function consisting of a sequence of constant-value segments of duration Δt. In more detail, we replace the varying value of $x(t)$ on the first interval $[0, \Delta t]$ with constant value x_0 (the value at the start of the interval), similarly with constant value x_1 on the second interval $[\Delta t, 2\Delta t]$, and so on

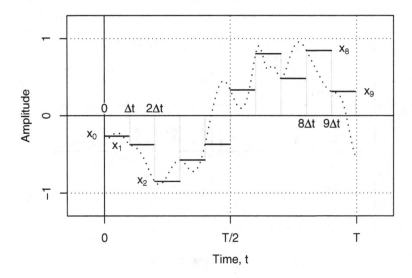

Figure 4.6 Piecewise approximation of a continuous function.

until the final segment with value x_{N-1} on the Nth interval $[(N-1)\Delta t, T]$. This is shown diagrammatically in Figure 4.6 for $N = 10$.

In Figure 4.6, the continuous function is shown by the dashed curve and the piecewise approximation by solid horizontal line segments, and there are two important observations we can make. First, it is important to note the sense of the sample-interval sequence. Each sample is *followed* by an interval Δt, *i.e.* we start the sampling at $t = 0$ and take the final sample at $t = T - \Delta t$, *not* at $t = T$. If we were to include a sample at $t = T$, this would effectively extend the duration to $T + \Delta t$ (or be the first sample in the next cycle of duration T). Second, this piecewise approximation has implications for features in the continuous time-function which vary faster than the sampling (frequency) can resolve and which are thus either lost from the discrete time-series or revealed as lower frequency features if there are cycles over more than one sample, which we will consider in Section 6.2.

Leaving such too-fast-feature issues aside for the present, we can substitute this piecewise linear approximation in place of $x(t)$ in Equation 4.21 and re-express the integral in terms of a summation over the N samples to obtain, in the first instance, a prototype forward DFT having the form, for the *kth* frequency coefficient

$$X_k = \frac{1}{T} \sum_{n=0}^{N-1} x_n e^{-2\pi i f_k t_n} \Delta t. \tag{4.24}$$

However, we can simplify this, noting that $f_k = k/T$ (k cycles over T), $t_n = n \,\Delta\, t$ and $T = N \,\Delta\, t$ to obtain

$$X_k = \frac{1}{N} \sum_{n=0}^{N-1} x_n e^{-2\pi i \frac{kn}{N}}. \qquad (4.25)$$

This is the basic forward DFT.

Remember: this is presented as an intuitively reasonable derivation which gives initial insights into some of the frequency resolution constraints of the DFT but is *not* the formal derivation of the DFT. There are some consequences of moving from an infinite continuous transform to a finite discrete transform that are important when using the DFT to identify periodic features in data which are not revealed by this approach which are considered in Chapter 6.

4.6 The DFT as a Linear Transformation

It is clear that the exponential term in Equation 4.25 is 2π-periodic and this means that both the time and frequency indexing, *i.e.* n and k, are modulo-N. To illustrate this, for a given value of n and values of k differing by integer multiples of N we have

$$e^{-2\pi i \frac{n(k+mN)}{N}} = e^{-2\pi i \frac{nk}{N}} e^{-2\pi i \frac{nmN}{N}}$$
$$= e^{-2\pi i \frac{nk}{N}} e^{-2\pi i mn} \qquad (4.26)$$

for integers $m = 0, \pm 1, \pm 2, \ldots$.

Noting that the product mn must be an integer, we have

$$e^{-2\pi i \frac{n(k+mN)}{N}} = e^{-2\pi i \frac{nk}{N}} \qquad (4.27)$$

meaning that $X_{k \pm mN} = X_k$ for all integers m.

Thus, we can deduce that there are only N distinct frequency components in the DFT spectrum and the index range $0 \le k \le N - 1$ contains all the DFT frequency information. Any other range of N consecutive integers contains shifted copies of the information in this 'central' spectrum. The full DFT spectrum is N-periodic: every time we move through N consecutive integers the central spectrum repeats. We return to this in the next section, and the periodic nature of the DFT spectrum is shown schematically in Figure 4.8.

4.6.1 The DFT Matrix and the Fast Fourier Transform

If we consider the discrete time and frequency functions as N-element vectors, then we can consider Equation 4.25 as representing the inner product of the

kth row of an $N \times N$ matrix (column-index n, row-index k) and the N-element time function. If we represent the time function as the N-element vector, $\underline{x}$, and extend the left-hand side of Equation 4.25 from a single frequency term, X_k, to the N-element vector, $\underline{X}$, *i.e.*

$$\underline{x} = \begin{pmatrix} x_0 \\ x_1 \\ \vdots \\ x_n \\ \vdots \\ x_{N-1} \end{pmatrix} \text{ and } \underline{X} = \begin{pmatrix} X_0 \\ X_1 \\ \vdots \\ X_k \\ \vdots \\ X_{N-1} \end{pmatrix} \quad (4.28)$$

then Equation 4.25 becomes

$$\underline{X} = \mathbf{F}\underline{x} \quad (4.29)$$

where $\mathbf{F}$ is the $N \times N$ DFT matrix.

In other words, the DFT can be considered as a linear transformation: more precisely as an orthogonal linear transformation because the elements in the frequency function are orthogonal. The DFT matrix is a complex symmetric matrix with general kth row corresponding to the coefficients in Equation 4.25 for $0 \le k \le N - 1$, *i.e.*

$$\mathbf{F} = \frac{1}{N} \begin{pmatrix} 1 & 1 & 1 & \cdots & 1 \\ 1 & e^{-2\pi i \frac{1}{N}} & e^{-2\pi i \frac{2}{N}} & \cdots & e^{-2\pi i \frac{N-1}{N}} \\ 1 & e^{-2\pi i \frac{2}{N}} & e^{-2\pi i \frac{4}{N}} & \cdots & e^{-2\pi i \frac{2(N-1)}{N}} \\ \vdots & & & & \vdots \\ 1 & e^{-2\pi i \frac{k}{N}} & e^{-2\pi i \frac{2k}{N}} & \cdots & e^{-2\pi i \frac{k(N-1)}{N}} \\ \vdots & & & & \vdots \\ 1 & e^{-2\pi i \frac{N-1}{N}} & e^{-2\pi i \frac{2(N-1)}{N}} & \cdots & e^{-2\pi i \frac{(N-1)^2}{N}} \end{pmatrix} \quad (4.30)$$

There are evident symmetries in this matrix and Fast Fourier Transform (FFT) algorithms exploit those symmetries to minimise the number of calculations required to produce the DFT of a time-series (or other discrete function). Appendix A presents an overview of how this basic form of DFT matrix varies according to N-even and N-odd, and presents specific DFT matrices for $N = 10$ and $N = 11$ to illustrate the basic symmetries of N-even and N-odd forms.

4.6.2 The Inverse DFT

In order for the DFT to be invertible, the matrix $\mathbf{F}$ must have an inverse matrix, $\mathbf{F}^{-1}$, such that $\mathbf{F}\mathbf{F}^{-1} = \mathbf{F}^{-1}\mathbf{F} = \mathbf{I}$, where $\mathbf{I}$ is the the $N \times N$ identity matrix. Therefore, if matrix $\mathbf{F}^{-1}$ exists then, from Equation 4.29 we must have

$$\mathbf{F}^{-1}\underline{X} = \mathbf{F}^{-1}\mathbf{F}\underline{x} = \underline{x}.$$

Inspection of the elements of matrix $\mathbf{F}$ indicates that its complex conjugate matrix, but without the scaling factor $1/N$, satisfies the requirements of the inverse matrix. Thus we have the inverse DFT matrix, also complex symmetric (row-index n, column-index k to match the indexing of x_n and X_k respectively), *i.e.*

$$\mathbf{F}^{-1} = \begin{pmatrix} 1 & 1 & 1 & \cdots & 1 \\ 1 & e^{2\pi i \frac{1}{N}} & e^{2\pi i \frac{2}{N}} & \cdots & e^{2\pi i \frac{N-1}{N}} \\ 1 & e^{2\pi i \frac{2}{N}} & e^{2\pi i \frac{4}{N}} & \cdots & e^{2\pi i \frac{2(N-1)}{N}} \\ \vdots & & & & \vdots \\ 1 & e^{2\pi i \frac{n}{N}} & e^{2\pi i \frac{2n}{N}} & \cdots & e^{2\pi i \frac{n(N-1)}{N}} \\ \vdots & & & & \vdots \\ 1 & e^{2\pi i \frac{N-1}{N}} & e^{2\pi i \frac{2(N-1)}{N}} & \cdots & e^{2\pi i \frac{(N-1)^2}{N}} \end{pmatrix} \quad (4.31)$$

Thus, the inverse DFT also exists and, in terms of a general element of the time function, x_n, we extract the coefficients from a general nth row and express the inner product explicitly, *i.e.*

$$x_n = \sum_{k=0}^{N-1} X_k e^{2\pi i \frac{kn}{N}}. \quad (4.32)$$

4.6.3 Linearity of the DFT

The linearity of the DFT follows directly from both the linearity of the (continuous-time) Fourier transform and the fact that the DFT is a linear transformation.

If we have two general discrete time functions (as N-element vectors) $\underline{x}$ and $\underline{x}'$, then their DFTs are

$$\underline{X} = \mathbf{F}\underline{x}$$
$$\underline{X}' = \mathbf{F}\underline{x}' \quad (4.33)$$

where $\mathbf{F}$ is the DFT matrix.

Therefore, if we have a discrete N-element time function $\underline{y} = A\underline{x} + B\underline{x}'$, where A and B are scalar multipliers, then the DFT of $\underline{y}$ is

$$\underline{Y} = \mathbf{F}\underline{y} = \mathbf{F}\left(A\underline{x} + B\underline{x}'\right)$$
$$= A\mathbf{F}\underline{x} + B\mathbf{F}\underline{x}' = A\underline{X} + B\underline{X}' \tag{4.34}$$

The inverse DFT is also linear.

4.7 The DFT Frequency Spectrum

Having established that there are N distinct frequency coefficients in the DFT spectrum and that there are complex-conjugate relationships between those coefficients (see Appendix A), we can deduce the form of the DFT spectrum. Summarising the complex-conjugate symmetries, we have:

1. For index $k = 0$ we have X_0, the zero-frequency (f_0) coefficient, the mean level, and it is pure real.
2. For index $k = 1$ we have X_1, the positive fundamental frequency (f_1) coefficient. However, for index $k = N - 1$ we have X_{N-1} which is the complex-conjugate of X_1 and so must be the negative fundamental frequency ($-f_1$ or f_{-1}) coefficient.
3. Similarly for other positive–negative frequency coefficient pairs. Counting in from the ends towards the centre according to the index, k, we have complex-conjugate pairs at X_2 and X_{N-2} ($\pm f_2$ or $f_{\pm 2}$ coefficients), X_3 and X_{N-3} ($\pm f_3$ or $f_{\pm 3}$ coefficients) and so on, approaching the centre ($k = N/2$).
4. At the centre, the behaviour is different for N-even and N-odd.

   i) For N-even, $N/2$ is an integer and so there is one central coefficient $X_{N/2}$ at $k = N/2$, the $N/2$ frequency coefficient. This is the coefficient of the highest harmonic, $f_{N/2}$, present in the DFT spectrum for N-even.
   ii) For N-odd, $N/2$ is not an integer and so there are two central coefficients $X_{(N-1)/2}$ and $X_{N-(N-1)/2}$, a complex-conjugate pair at $k = N/2 \pm 1/2$ (noting $(N+1)/2 = N - (N-1)/2$). These are coefficients for the highest harmonic, $\pm f_{(N-1)/2}$ (or $f_{\pm(N-1)/2}$), present in the DFT spectrum for N-odd.

We can combine the information in the preceding sections to produce a summary of the frequency indexing in terms of variable m, $0 \le m \le M$, which can be considered as the index of the harmonic frequencies of the DFT spectrum, compared to k which is effectively a positional index in the DFT spectrum. Thus, we have

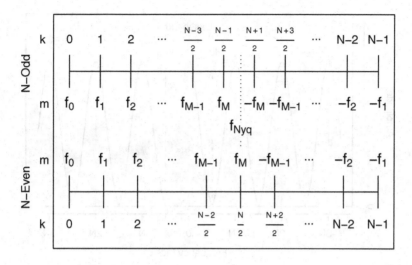

Figure 4.7 Indexing in N-even and N-odd DFT spectra.

$$M = \begin{cases} \frac{N}{2}, & N \text{ even} \\ \frac{N-1}{2}, & N \text{ odd} \end{cases}. \qquad (4.35)$$

This enables us to conveniently summarise the general form of the DFT spectrum, showing the relationship between the position, k, and the order of the harmonic, m, as shown in Figure 4.7 (the frequencies are labelled $\pm f_m$ rather than $f_{\pm m}$ for clarity).

It is clear that the maximum resolvable frequency is lower for N-odd $\pm f_{(N-1)/2}$ ($f_{\pm(N-1)/2}$) than N-even ($f_{N/2}$). In terms of the number of samples, N, for a time-series, the highest possible resolvable frequency is $f_{N/2}$, $i.e.$ $N/2$ cycles over the duration $T = N\Delta t$. This frequency is referred to as the Nyquist frequency after H Nyquist (1889–1976), and is shown in Figure 4.7 as f_{Nyq}.

The DFT spectrum is asymmetric: $i.e.$ it does not start at $-f_{\text{Nyq}}$, with f_0 in the centre, and end at f_{Nyq}. Instead, it starts at f_0, then increases through the positive frequencies towards f_{Nyq}, then decreases through the negative frequencies to f_{-1}. This is a consequence of the indexing range, $0 \le k \le N-1$, which is itself asymmetric with regard to the order of the harmonics. If we were to index on the range $-N/2 \le k \le N/2$ then the DFT spectrum would be symmetric ($i.e.$ $\pm f_{\text{Nyq}}$ at the ends with f_0 in the centre), but this would be inconvenient for N-odd. The relationship between the 'central' DFT spectrum as indexed on $0 \le k \le N-1$ and the 'baseband' spectrum $-f_{\text{Nyq}} \le f \le f_{\text{Nyq}}$ is shown schematically in Figure 4.8, which also shows the cyclically repeating nature of the DFT spectrum.

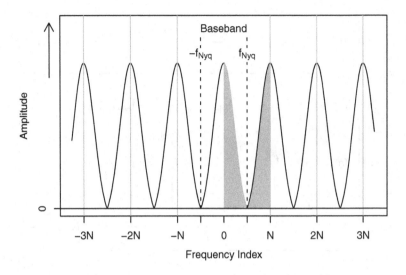

Figure 4.8 Central and baseband DFT spectrum.

In Figure 4.8, the central spectrum indexed on $0 \leq k \leq N - 1$ is shown shaded. The baseband spectrum on $-f_{\text{Nyq}} \leq f \leq f_{\text{Nyq}}$ (notional indexing $-N/2 \leq k \leq N/2$) is shown by vertical dashed lines. The positive-frequency portion of the baseband spectrum is contained in the left-hand (lower) half of the central spectrum; the negative-frequency portion of the baseband spectrum is contained in the right-hand (upper) half of the central spectrum.

4.8 Nyquist–Shannon Sampling Theorem

These maximum-frequency considerations are formalised by the Nyquist–Shannon sampling theorem, often referred to as 'Shannon's sampling theorem' or just 'sampling theorem', which was formalised by C E Shannon (1949) following work by others, principally H Nyquist (1928).

For a time-series of duration $T = N\Delta t$ with constant sampling interval Δt and constant sampling frequency $f_s = 1/\Delta t$, this theorem can be expressed as:

- The maximum frequency which can be resolved without loss of information in a discrete function sampled with constant sampling frequency f_s is the Nyquist frequency, $f_{\text{Nyq}} = f_s/2$, i.e. half the sampling frequency.

This is important: there are no circumstances in (equal-interval) discrete/sampled data under which we can correctly resolve a frequency higher than half the sampling frequency and, furthermore, there are some circumstances under which we cannot correctly resolve a frequency equal to half the sampling frequency. This constraint is not unique to the DFT: no other techniques can correctly and unambiguously resolve frequencies higher than half the sampling frequency in sampled data.

Also, do not confuse the Nyquist frequency with the Nyquist rate: the Nyquist rate is the minimum sampling frequency required to 'see' a given maximum frequency in sampled data.

4.8.1 Resolving the Nyquist Frequency

We deduced from Section 4.5 and Figure 4.6 that the sampling and DFT are unable to 'see' features which vary too fast in relation to the sampling frequency and we have established that the maximum frequency that sampling and the DFT can see is half the sampling frequency. However in the preceding section, we also observed the special case of coefficient $X_{N/2}$ being pure real and this has implications for resolving features at $f_{Nyq} = f_s/2$.

This is illustrated in Figure 4.9 which shows the simple case of one period of an arbitrary waveform of fundamental frequency f_1 and arbitrary phase (grey line), with waveform plus cosine and sine at the Nyquist frequency shown by solid and dashed lines respectively. The waveform is sampled at $f_s = 10f_1$, which means that the Nyquist frequency $f_{Nyq} = 5f_1$.

In Figure 4.9 it is clear that the samples cannot distinguish the waveform-plus-sine from the underlying waveform because the two waveforms intersect

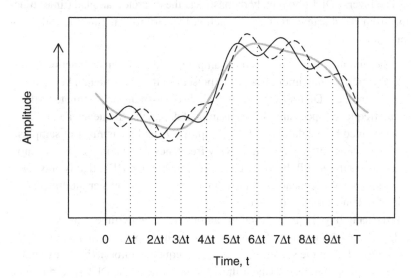

Figure 4.9 Sinusoid with fifth harmonic sine (dashed) and cosine (solid).

at the sampling points. However, it is clear that the samples can distinguish the waveform-plus-cosine from the underlying waveform because the two waveforms do not intersect at the sampling points (intersect mid-way between sampling points). Therefore, we can deduce directly from Figure 4.9 that the sampling and DFT are unable to 'see' a sine at the Nyquist frequency but are able to 'see' a cosine at the Nyquist frequency.

More generally, the sampling and DFT are able to 'see' a sinusoid with frequency equal to the Nyquist frequency to an extent determined by its phase with respect to the start of the time-series, $t = 0$. However, the sampling and DFT are unable to correctly 'see' sinusoids with frequencies higher than the Nyquist frequency under any circumstances (see Section 6.2).

4.9 Summary

The *discrete* Fourier transform, DFT of a *discrete* function in the time domain is a frequency domain representation of it, *i.e.* a *discrete* function in the frequency domain. DFTs are not restricted to time–frequency relationships: the DFT of a discrete function in an arbitrary domain is a discrete function in the corresponding frequency domain. For example, the Fourier transform of a discrete spatial function (*e.g.* with distance units of metres) will be a discrete spatial-frequency function with frequencies in terms of cycles per unit distance (*e.g.* cycles per metre).

The inverse DFT completely reconstructs the discrete, sampled (time) function from the discrete frequency function (spectrum) – noting the following conditions.

- If the sampling frequency used to sample the discrete (time) function is at least twice the maximum frequency present in the phenomenon being monitored, the DFT spectrum will be a discrete-frequency estimate of the true frequency spectrum of the unsampled data, with frequency resolution determined by the duration of the sampling, T, and the number of samples, N. This does not guarantee that every frequency present in the phenomenon being monitored will be precisely represented by the DFT: that is only the case where all frequencies in the data are harmonics (integer multiples) of the fundamental DFT frequency.
- If the sampling frequency used to sample the discrete (time) function is less than twice the maximum frequency present in the phenomenon being monitored, then the sampling loses high-frequency information above half the sampling frequency. Under these circumstances, the DFT spectrum will

be a (distorted, 'aliased') discrete-frequency estimate of the frequency spectrum below half the sampling frequency in the sampled data, with frequency resolution determined by the duration of the sampling, T, and the number of samples, N, but will only partially represent the true frequency spectrum of the unsampled data as information has already been lost in the sampling process.

To summarise, the DFT spectrum of an N-point time-series of duration T consists of :

- coefficient of zero-frequency term representing the mean of the time-series, sometimes referred to as the 'dc' (direct current) term;
- coefficients of frequency components in complex-conjugate pairs at $\pm f_1$ ($f_{\pm 1}$), $\pm f_2$ ($f_{\pm 2}$), and so on up to the maximum resolvable frequency, with frequency interval $\Delta f = |f_1| = 1/T$;

    – a single pure-real frequency coefficient for maximum frequency $f_{N/2}$ for N-even;
    – a complex-conjugate pair of frequency coefficients for maximum frequency $\pm f_{(N-1)/2}$ ($f_{\pm(N-1)/2}$) for N-odd.

Exercises

1. Write down the forward and inverse DFT matrices for $N = 3$, using the skeleton forms in Section 4.6 for guidance, and confirm by explicitly multiplying the two matrices together that one inverts the other.
2. For $N = 10$ and the general formula $ssoid = cos(2\pi f \frac{tb}{10}) + sin(2\pi f \frac{tb}{10})$ use the following commands to generate a sinusoid ssoid of frequency 1 cycle over time-base tb.

```
tb <- 0:9
ssoid <- cos(2*pi*1*tb/10) + sin(2*pi*1*tb/10)
```

Then FFT that sinusoid and confirm that the spectrum consists of only two non-zero coefficients, *i.e.* the $f_{\pm 1}$ coefficients, and that they are a complex-conjugate pair.
3. Repeat Exercise 2 for frequencies $2, 3, 4$ cycles over tb, and confirm that the spectra consist of only two non-zero coefficients, *i.e.* the $f_{\pm 2}, f_{\pm 3}, f_{\pm 4}$ coefficients respectively, and that they are complex-conjugate pairs at each frequency.

4. Repeat Exercise 2 for the frequency 5 cycles over tb, and confirm that the
   spectrum consists of only one non-zero coefficient, *i.e.* the f_5 coefficient,
   and that it is pure real.

5. For $N = 11$ and the general formula $ssoid = cos(2\pi f \frac{tb}{11}) + sin(2\pi f \frac{tb}{11})$
   use the following commands to generate a sinusoid ssoid of frequency 1
   cycle over time-base tb.

```
tb <-  0:10
ssoid <- cos(2*pi*1*tb/11)  +  sin(2*pi*1*tb/11)
```

Then FFT that sinusoid and confirm that the spectrum consists of only two
non-zero coefficients, *i.e.* the $f_{\pm 1}$ coefficients, and that they are a
complex-conjugate pair.

6. Repeat Exercise 5 for frequencies 2, 3, 4, 5 cycles over tb, and confirm that
   the spectra consist of only two non-zero coefficients, *i.e.* the
   $f_{\pm 2}, f_{\pm 3}, f_{\pm 4}, f_{\pm 5}$ coefficients respectively, and that they are
   complex-conjugate pairs at each frequency.

5

Using the FFT to Identify Periodic Features in Time-Series

In Chapter 4 we noted that Fast Fourier Transform (FFT) is the name used to refer to highly optimised algorithms to calculate the discrete Fourier transform (DFT). However, to most intents and purposes, both names and abbreviations are interchangeable: indeed, since the development of FFT algorithms it would be unlikely that something described as a DFT was not calculated using an FFT algorithm of some kind.

From this point on, having established the essential underpinning theory for understanding the DFT output spectrum in the context of the analysis of periodic features in time-series, we will routinely use the term FFT unless this could result in a loss of accuracy or clarity. There is a little more theory to come with regard to what the FFT (DFT) cannot do but, before that, let us look at what it can do. This chapter presents examples of use of the FFT, starting with straightforward and progressing to more complicated spectra. All the datasets used are contained in R standard or add-on library packages.

5.1 The Standard FFT

In many software packages, including R, the DFT normalising factor of $1/N$ is dropped, because we generally need only the relative magnitudes of $\{X_k\}$ and not the normalised ones. Thus, we have a *de facto* standard form for the forward and inverse transform pair. The forward transform, from time to frequency ($X = \mathcal{F}(x)$, with $-i$ in the complex exponential), and the inverse transform, from frequency to time ($x = \mathcal{F}^{-1}(X)$, with $+i$ in the complex exponential), are

$$X_k = \sum_{n=0}^{N-1} x_n e^{-2\pi i \frac{kn}{N}}$$

$$x_n = \sum_{k=0}^{N-1} X_k e^{2\pi i \frac{kn}{N}} \qquad (5.1)$$

respectively, for discrete time function $x(t) = x_0, x_1, x_2, \ldots, x_{N-1}$ with $x_n = x(t_n)$, and discrete frequency spectrum $X(f) = X_0, X_1, X_2, \ldots, X_{N-1}$ with $X_k = X(f_k)$.

This is the form of the FFT (DFT) used throughout the subsequent material of the book.

5.1.1 Other Implementations of the FFT

In some software packages, the normalising factor of $1/N$ is included in the forward transform or, sometimes, in the inverse transform. It is also possible that you might encounter an implementation which includes a factor of $1/\sqrt{N}$ in both forward and inverse transforms.

If you are unsure with regard to the FFT-implementation in any software that you are using then consult the documentation or other *reliable* source of information. Alternatively, the first element in the FFT output array, X_0, corresponds to the mean value of the data (sometimes referred to as the 'dc level'). Therefore, a reliable way of checking whether a forward transform is normalised, for real data (assuming that the mean value is non-zero), is to divide X_0 by the mean value of the data. If this yields values of N, $\sqrt{N}$ or 1, for example, then the forward FFT implementation you are using is either unnormalised or normalised by $\sqrt{N}$ or N respectively. Corresponding considerations apply to the inverse FFT implementation.

Note that if the software you are using implements the FFT in an unnormalised form (as R does) then you will have to explicitly take account of the normalising factor yourself as appropriate in any analysis you are undertaking, and is something we have to do in Chapter 9 when considering variance.

5.2 Time-Series with Few Frequencies

For a first hands-on example, let us investigate the FFT of a time-series that will give us a clear and unambiguous spectrum. This is the nottem dataset which we used in Chapter 2.

5.2.1 Preliminary Investigation

It is always advisable to do some preliminary investigation before analysing the FFT spectrum: this can give you useful insights as to what to investigate in more detail and help you to avoid missing something of interest or misinterpreting the FFT spectrum. In the context of this 240-sample 20-year temperature record, a line-plot as in Figure 2.10, is a reasonable starting point and this gives us a clear expectation of an FFT spectrum showing frequency 'markers' corresponding to the strong annual cycle.

5.2.2 Frequency Spectrum, FFT

To perform the FFT and inverse FFT on this dataset, enter the following two commands.

```
nottemf <- fft(as.numeric(nottem))
nottemfi <- fft(nottemf, inverse=TRUE)
```

The first command performs the FFT (the as.numeric() command strips the embedded date information, not strictly necessary) and writes the output to a new object called nottemf which is a vector of 240 complex numbers. The second command performs the inverse FFT and writes the output to a new object called nottemfi which in theory should be a vector of 240 real numbers, each number equal to 240 times the corresponding number in nottem. In practice, owing to the precision to which R can do the calculations and the rounding and storage errors in the numbers, nottemfi will be vector of 240 complex numbers in which all the real parts are 240 times the corresponding number in nottem and the imaginary parts are zero or insignificantly different from zero.

Note on R and variable names

The choice of nottemf and nottemfi as variable names is entirely arbitrary – you can name these variables as you wish. More generally, it is sensible to choose variable names that help you to keep track of things – and be careful not to overwrite variables by accidentally reusing variable names.

Basic checks

The following four commands perform some basic checks on the FFT output.

```
nottemf[1]
Re(nottemf[1])/mean(nottem)
summary(Im(nottemfi))
summary(Re(nottemfi)/nottem)
```

The four commands print to the console as follows.

```
> nottemf[1]
[1] 11769.5+0i
> Re(nottemf[1])/mean(nottem)
[1] 240
> summary(Im(nottemfi))
      Min.    1st Qu.     Median       Mean    3rd Qu.       Max.
-2.189e-12 -3.768e-13  1.229e-13  0.000e+00  4.639e-13  1.775e-12
> summary(Re(nottemfi)/nottem)
  Min. 1st Qu.  Median    Mean 3rd Qu.    Max.
   240     240     240     240     240     240
```

The first command prints the zero-frequency coefficient, X_0, to the console and this number is pure real although stored and listed as a complex number. The second command divides the real part of X_0 by the mean of the data and prints this number to the console, confirming that R does not include a normalising factor in the forward FFT. The third command prints a summary (*i.e.* minimum, first quartile, median, mean, third quartile and maximum values) of the imaginary part of the FFT output to the console and the six values are (indistinguishable from) zero. The fourth command prints a summary of the ratios between corresponding values in nottemfi and nottem and confirms the ratio is constant.

More detailed checks of complex conjugate symmetry

The time-series nottem has length 240, which is even, and we have already confirmed that the first coefficient, nottemf[1], is pure real. In the remaining 239 coefficients, nottemf[2:240], we are expecting that the first 119, nottemf[2:120], and the last 119, nottemf[122:240], are complex numbers with the last 119 being the complex conjugates of the first 119 according to the frequency indexing we discussed in Chapter 4. We are also expecting that the middle coefficient, nottemf[121], is pure real. To quickly check and confirm this, enter the following four commands.

```
c(nottemf[2], nottemf[240])
c(nottemf[3], nottemf[239])
c(nottemf[120], nottemf[122])
nottemf[121]
```

The four commands print to the console as follows.

```
> c(nottemf[2], nottemf[240])
[1] 30.42129+73.46971i 30.42129-73.46971i
> c(nottemf[3], nottemf[239])
[1] -10.55785+11.1371i -10.55785-11.1371i
> c(nottemf[120], nottemf[122])
[1] 7.231881-1.26345i 7.231881+1.26345i
> nottemf[121]
[1] 46.9+0i
```

The first, second and third commands respectively print the pairs of complex-conjugate frequency coefficients X_1 & X_{-1} , X_2 & X_{-2} and X_{119} & X_{-119} to the console, and these clearly show the complex conjugate nature of the positive and negative frequency coefficients. The fourth command prints the single real maximum frequency coefficient X_{120} to the console.

5.2.3 Amplitude Spectrum

An informative type of plot is a quasi-histogram bar-plot (column-plot) of the magnitudes of the FFT coefficients, for all or part of the frequency spectrum depending on context. Instead of bar-plots, it is possible to use line plots, which are reasonably commonly encountered, or point or other plot-types depending on the context.

The following commands produce a basic frequency coefficient index and a very basic plot of the magnitudes of the coefficients, with default axis labels, as shown in Figure 5.1.

```
f.ind <- 0:239
plot(f.ind, abs(nottemf), type="h")
```

This plot reveals a large frequency marker at f_0, and distinct symmetric frequency markers at 20 frequency intervals in from the positive and negative ends of the spectrum plus small but still visible symmetric frequency markers at a further 20 intervals in towards the centre.

To improve this plot, *e.g.* with informative axes and labels, we need to scope the data and confirm the positions of the symmetric pairs of frequency markers. It is also helpful to start to develop a systematic approach to generating frequency (harmonic) indices. Therefore, in the following three commands, the first generates an index for positive and negative frequencies in terms of

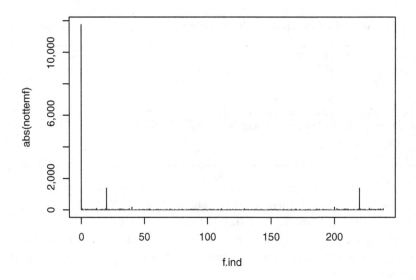

Figure 5.1 FFT amplitude spectrum, nottem time-series.

cycles over the whole period, the second binds the frequency index and the FFT coefficients into a data-frame nottemfop and the third generates a column of FFT magnitudes nottemfop$mag.

```
f.ind <- c(seq(0,240/2,1), seq(-(240/2-1),-1,1))
nottemfop <- data.frame(f.ind, nottemf)
nottemfop$mag <- abs(nottemfop$nottemf)
```

We can then sort the data-frame in descending order of the coefficient magnitudes and print the largest nine (sufficient in this example) to the console to scope the spectrum, as in the following two commands.

```
nottemfops <- nottemfop[order(nottemfop$mag,
decreasing=TRUE),]
nottemfops[1:9,]
```

This reveals that X_0 coefficient, is the largest; followed by the $X_{\pm 20}$ coefficients (20 cycles in 20 years, *i.e.* 1 cycle/year); followed by the $X_{\pm 40}$ coefficients (40 cycles in 20 years, *i.e.* 2 cycles/year), with none of the other frequency markers (coefficients) being significant.

	f.ind	nottemf	mag
1	0	11769.50000+ 0.000000i	11769.50000
21	20	-1376.79904+166.86479i	1386.87399
221	-20	-1376.79904-166.86479i	1386.87399
41	40	150.85000- 98.29388i	180.04835
201	-40	150.85000+ 98.29388i	180.04835
13	12	95.72705+ 26.35122i	99.28774
229	-12	95.72705- 26.35122i	99.28774
39	38	-73.14099+ 43.995523i	85.35329
203	-38	-73.14099- 43.995523i	85.35329

Now that we have a 'proper' frequency harmonic index, we can produce a calibrated frequency index in terms of cycles per year and plot a symmetric amplitude spectrum as shown in Figure 5.2 as follows.

```
nottemfop$f.cal <- nottemfop$f.ind/20
plot(nottemfop$f.cal, nottemfop$mag, type="h",
xlim=c(-6,6), xaxp=c(-6,6,12), xlab="Frequency,
cycles/y", ylab="Amplitude")
```

The amplitude spectrum in this sort of calibrated form, either asymmetric or symmetric, is useful for indicating the relative magnitudes of the periodic

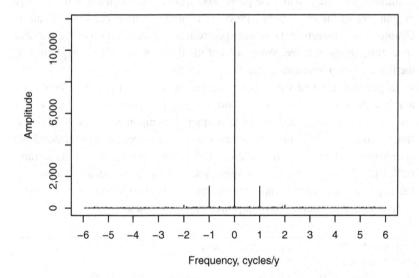

Figure 5.2 FFT amplitude spectrum, nottem time-series.

features and also the magnitudes of the periodic features relative to the mean ('dc') level. However, the amplitude spectrum does not directly indicate effect-size and, for that, we need the so-called power spectrum, as described in the next section. Also, it is often the case that the mean level is neither helpful nor informative. This is the case where the measurement datum is arbitrary, as it is with temperature in this example, where we are interested in the cyclic variation. If we were to convert these temperatures to degrees Celsius and FFT the converted time-series, then the relative magnitudes of the non-zero-frequency coefficients would be the same but the relative magnitude of the zero-frequency X_0 coefficient to the non-zero-frequency coefficients would be different. We will consider this in more detail in subsequent examples.

5.2.4 Power Spectrum

The terminology 'power spectrum' arises from the use of the Fourier transform in signal analysis and related fields to determine how transmitted power is distributed across the frequencies in, for example, an electric current or electromagnetic signal, and power is proportional to the square of the amplitude. In the context of time-series analysis, power has no general meaning but the terminology is retained and the power spectrum is the spectrum of the squared magnitudes of the coefficients. Power is related to variance, as we discuss in Chapter 9, and therefore the power spectrum is a representation of the variance spectrum, *i.e.* the relative proportions of the time-series variance explained by the frequency components in the FFT spectrum.

In general, for real time-series, we are interested in the total power (or amplitude) at a given frequency and are not interested in the symmetrical division over positive and negative frequency components. In the following three commands, the first produces the power spectrum nottemfop$pwr at positive and negative frequencies, the second creates a new data-frame nottemfop.t for the positive frequencies and the third adds the power at the negative frequency components to the corresponding positive frequency components to conserve the total power in the spectrum.

```
nottemfop$pwr <- nottemfop$mag^2
nottemfop.t <- nottemfop[1:121,]
nottemfop.t$pwr[2:120] <- nottemfop$pwr[2:120] +
nottemfop$pwr[240:122]
—— or ——
nottemfop.t$pwr[2:120] <- 2*nottemfop.t$pwr[2:120]
```

With regard to the first command, we could also calculate the power by multiplying the complex coefficients by their complex conjugates. With regard to the third command, we could make use of the symmetries of the FFT output spectrum and, instead of explicitly adding the powers at the positive and negative frequencies for $|k| < N/2$, we could have multiplied the positive-frequency powers for $1 \le k < N/2$ by 2. This is shown in the fourth command (remembering that for N-even there is only a single positive-frequency power at $k = N/2$ which, therefore, is not multiplied by 2). As multiplying positive-frequency powers (or amplitudes) by 2 is an easier and more convenient approach, this is the procedure we will follow in further examples, but remember that this short-cut is only applicable for real-valued data.

The following two commands plot this power spectrum with a logarithmic vertical axis.

```
plot(nottemfop.t$f.cal, nottemfop.t$pwr, type="h",
xlab="Frequency, cycles/y", ylab="Power, arb. units",
log="y", yaxt="no")
axis(2, at=c(1e02, 1e03, 1e04, 1e05, 1e06, 1e07, 1e08),
labels= c(expression(paste("10"^"2")), "",
expression(paste("10"^"4")), "",
expression(paste("10"^"6")), "", expression
(paste("10"^"8"))))
```

This produces a plot as in Figure 5.3. Note the logarithmic vertical axis and the relative magnitudes of the markers at 0, 1 and 2 cycles/year compared to the other markers. The magnitude of the marker at 0 cycles/year is *ca.* 36 times the magnitude of the marker at 1 cycle/year; the magnitude of the marker at 1 cycle/year is *ca.* 59 times the magnitude of the marker at 2 cycles/year; and the magnitude of the marker at 2 cycles/year is *ca.* 25 times the average magnitude of the other markers (*i.e.* all the non-zero-frequency markers except those at 1 and 2 cycles/year). These other frequencies can be considered as 'noise' in this case, and we overview noise in Chapter 8.

5.2.5 Further Comments

The nottem data reveal a clear annual cycle in the recorded temperatures, which is closely approximately annual sinusoidal with a small but visibly significant second harmonic. We would see exactly the same relationships between the magnitudes of the non-zero frequencies if we were to convert from degrees Fahrenheit to degrees Celsius or, indeed, any other temperature scale. However, if we were to convert from degrees Fahrenheit to degrees Celsius,

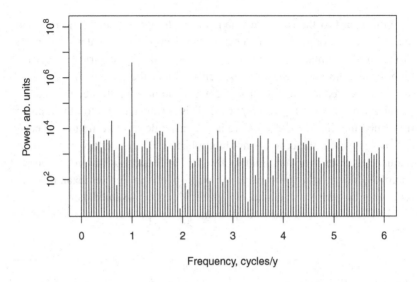

Figure 5.3 FFT power spectrum, nottem time-series.

for example, then the relationships between the magnitudes of the non-zero frequencies and the zero-frequency/mean level would change because of the relationship between the two temperature scales, *i.e.* $T_F = 1.8 \times T_C + 32$ and $0°F \neq 0°C$.

5.3 Time-Series with Many Frequencies

The nottem dataset provides a straightforward example with effectively a single-frequency spectrum. However, in practice, time-series and their FFT spectra are generally more complicated. As an example of a multi-frequency time-series, we will consider the Hillarys tide-height dataset in the R library package TideHarmonics.

The time-series Hillarys comprises 26,304 hourly sea-level readings at Hillarys, near Perth in SW Australia recorded over the three-year period 2012–2014. This time-series is stored as a two-column data-frame, column-1 is the date-time and column-2 is the sea-level in metres relative to the tide-gauge zero (which is an arbitrary reference datum).

5.3.1 Initial Investigation

First, we will produce a basic plot of the data as shown in Figure 5.4, as follows.

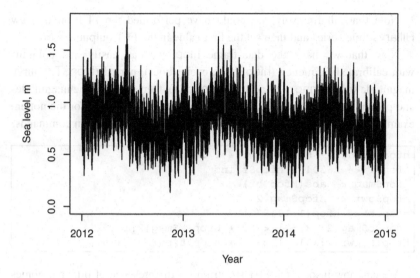

Figure 5.4 The Hillarys time-series.

```
plot(Hillarys$DateTime, Hillarys$SeaLevel, type="l",
xlab="Date", ylab="Sea Level, m")
```

Figure 5.4 shows the sea level relative to the tide-gauge zero and shows some very fast (at this time-scale) variations which correspond to the diurnal (*ca.* 24.8 hour) and semi-diurnal (*ca.* 12.4 hour) tides, plus some longer-period variations.

We can now proceed to investigate the time-series for the presence of tidal-harmonic frequencies.

5.3.2 Power Spectrum

We start by subtracting the mean sea-level, which subtracts the arbitrary tide-gauge datum, to obtain the time-series as variations around the mean sea level at Hillarys, and then FFT this time-series as in the first two commands below. As we did in the nottem example, we create a basic FFT frequency (harmonic) index (N-even) and bind this and the FFT output into a data-frame to simplify the subsequent analysis.

```
hy <- Hillarys$SeaLevel - mean(Hillarys$SeaLevel)
hf <- fft(hy)
f.ind <- c(seq(0,26304/2,1), seq(-(26304/2-1),-1,1))
hfop <- data.frame(f.ind, hf)
```

Note that, alternatively, we could have performed the FFT on the raw Hillarys time-series and then set the first value in the FFT output to zero.

Now that we have the data-frame hfop, we can easily add columns with calibrated frequency (hfop$f.cal, cycles/hour), period (hfop$T, hours), magnitude (hfop$mag) and power (hfop$pwr), then combine the magnitudes and powers at the positive and negative frequencies, as we did for the nottem example, into a new data-frame, hfop.t, using the following seven commands.

```
hfop$f.cal <- hfop$f.ind * 1/26304
hfop$T <- 26304 / hfop$f.ind
hfop$mag <- abs(hfop$hf)
hfop$pwr <- hfop$mag^2
hfop.t <- hfop[1:13153,]
hfop.t$mag[2:13152] <- 2 * hfop.t$mag[2:13152]
hfop.t$pwr[2:13152] <- 2 * hfop.t$pwr[2:13152]
```

We can now inspect the FFT spectrum for the presence of tidal harmonics and we start by plotting a basic power spectrum, as shown in Figure 5.5 with the following commands.

```
plot(hfop.t$f.cal, hfop.t$pwr, type="h", xlim=c(0,0.5),
xaxt="no", xlab="Frequency, cycles/h", ylab="Power, arb.
units")
axis(1, c(0, 0.1, 0.2, 0.3, 0.4, 0.5), c("0", "0.1",
"0.2", "0.3", "0.4", "0.5"))
axis(1, c(0.02, 0.04, 0.06, 0.08), labels=NA)
```

From Figure 5.5 it is clear that there is some low-frequency information, then further information at frequencies of *ca.* 0.04 and 0.08 cycles/hour, *i.e.* approximately diurnal and semi-diurnal cycles.

To investigate further, we will truncate the x-axis at $f = 0.09$ cycles/hour and expand the detail around the approximately diurnal and semi-diurnal cycles as shown in Figure 5.6, as follows.

```
plot(hfop.t$f.cal, hfop.t$pwr, type="h",
xlim=c(0.03,0.09), xaxt="no", xlab="Frequency,
cycles/h", ylab="Power, arb. units")
axis(1, c(0.03, 0.04, 0.05, 0.06, 0.07, 0.08, 0.09),
c("0.03", "0.04", "0.05", "0.06", "0.07", "0.08",
"0.09"))
```

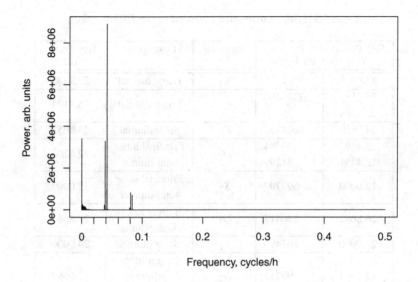

Figure 5.5 Power spectrum, Hillarys time-series.

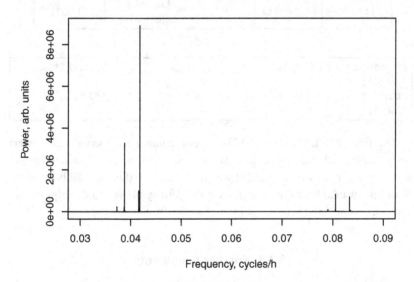

Figure 5.6 Power spectrum, Hillarys time-series, detail.

To resolve the frequencies and periods more accurately, we can select rows and columns and print these as lists to the R console as follows, for inspection and collation.

Table 5.1. *Tidal harmonics revealed in the Hillarys data*

FFT Period (h)	Power (arb. units)	Harmonic	Description	True Period (h)
23.9345	8901731	K_1	Lunar diurnal	23.9345
25.8135	3286494	O_1	Lunar diurnal	25.8193
25.8389	226966			
24.0659	1004339	P_1	Solar diurnal	24.0659
12.4193	823084	M_2	Principal lunar semi-diurnal	12.4206
12.4251	81299			
12.0000	699707	S_2	Principal solar semi-diurnal	12.0000
26.8682	236414	Q_1	Larger lunar elliptic diurnal	26.8684
24.0000	167499	S_1	Solar diurnal	24.0000
12.6583	93713	N_2	Larger lunar elliptic semi-diurnal	12.6583
11.9672	46330	K_2	Lunisolar semi-diurnal	11.9672

```
cbind(hfop.t[hfop.t$T>11.75 & hfop.t$T<12.75, c("T",
"pwr")])
cbind(hfop.t[hfop.t$T>23.5 & hfop.t$T<27, c("T",
"pwr")])
```

The first command prints a 175-row two-column list (period and power) which includes the semi-diurnal tidal harmonics. The second command prints a similar 145-row two-column list which includes the diurnal tidal harmonics. Both sets of tidal harmonics, with their standard symbols and descriptors, are summarised in Table 5.1.

5.3.3 Further Comments

The plots, lists and table indicate that some of the tidal harmonics, particularly the M_2 and O_1 harmonics, are revealed in the FFT spectrum as 'blurred' markers over several adjacent frequencies. This is because these frequencies are not integer multiples of the fundamental frequency in the FFT spectrum (1 cycle over 26.304 hours), *i.e.* are not FFT-harmonic frequencies. Any periodic component in the data that is not at an FFT-harmonic frequency will not be orthogonal to the FFT-harmonic sinusoids and so will be linearly dependent on the sinusoids in the FFT spectrum, as outlined in principle in Section 4.9. Therefore, rather than a single-frequency marker in the FFT output, such periodic components will result in a 'blur' of markers centred on the frequency of the periodic feature in question. This is discussed further in Chapter 6.

5.4 Time-Series with Trends

In the previous two examples, we have implicitly assumed that the residual data, when stripped of all the periodically varying components, is a uniform mean ('dc') level. This is not always the case: some time-series have trends, or trend-like features, and in this example we look at detrending a time-series prior to applying the FFT and interpreting the spectrum.

As an example, we will consider the (lowercase) co2 dataset included in the R datasets package. The co2 dataset is a 468-sample 39-year time-series (1959–1997, *i.e.* mid-January 1959 to mid-December 1997) of monthly atmospheric CO_2 concentration, in ppm, from Mauna Loa. (Note, not the CO2 dataset, that is something different.)

5.4.1 Initial Investigation

We will start by plotting the time-series to produce the plot as in Figure 5.6, plus the best-fit linear estimate of the trend in the time-series, as follows.

```
tb <- seq(1959+1/24, 1998-1/24, 1/12)
co2lm <- lm(co2 ~ tb)
plot(co2, xlab="Year", ylab=expression(paste("CO"[2]," 
atmos. conc'n, ppm")))
abline(co2lm, lty=2)
```

The first command generates a mid-monthly sequence of time-stamps for the linear regression in the second command. The third command plots the time-series and the fourth adds the regression line (dashed).

Figure 5.7 shows apparent annual variation around an approximately linear increasing trend but it is clear that the trend is not exactly linear as the regression line does not pass through the mid-points of all the annual cycles. In the first instance, we will follow the same procedure as in the previous section to produce a power spectrum, using the following commands. For this time-series, the basic unit of time is the year (subdivided into 12 months), and so

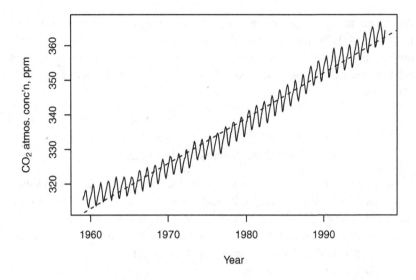

Figure 5.7 Raw co2 time-series.

we calibrate the frequency and period variables (co2op$f.cal and co2op$T respectively) in terms of years. The final command below produces a plot as shown in Figure 5.8.

```
f.ind <- c(seq(0,468/2,1), seq(-(468/2-1),-1,1))
co2f <- fft(co2)
co2op <- data.frame(f.ind, co2f)
co2op$f.cal <- co2op$f.ind * 1/39
co2op$T <- 39 / co2op$f.ind
co2op$mag <- abs(co2op$co2f)
co2op$pwr <- co2op$mag^2
co2op.t <- co2op[1:235,]
co2op.t$mag[2:234] <- 2 * co2op.t$mag[2:234]
co2op.t$pwr[2:234] <- 2 * co2op.t$pwr[2:234]
plot(co2op.t$f.cal, co2op.t$pwr, type="h", log="y",
ylim=c(1e3,1e11), xlab="Frequency, cycles/y",
ylab="Power, arb. units")
```

Figure 5.8 shows evidence of annual cycles, with some evidence of second and third harmonics, but this annual-plus-harmonic cycle information is obscured by noise and noise-like features, which decrease with frequency. These features arise in large part from the trend, because the FFT decomposes the trend mainly in terms of low-frequency sinusoids.

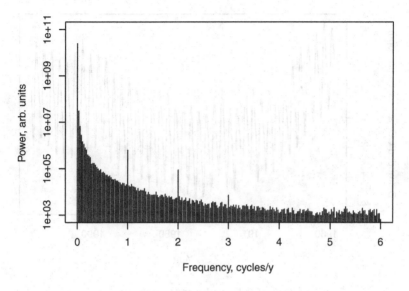

Figure 5.8 Power spectrum, raw co2 time-series.

5.4.2 Detrending the Data

For a more informative analysis, we need to detrend the time-series and we will use linear regression to predict the trend: in essence, we will subtract the values indicated by the dashed line in Figure 5.7 from the data. In the following three commands, the first extracts the linear trend from the regression output co2lm, the second subtracts that trend from the data to produce the detrended time-series, the third command plots the detrended time-series as shown in Figure 5.9.

```
co2.lin.trend <- fitted(co2lm)
co2.lin <- co2 - co2.lin.trend
plot(co2.lin, xlab="Year",
ylab=expression(paste("Detrended CO"[2]," atmos. conc'n,
ppm")))
```

Figure 5.9 shows the apparent annual variation much more clearly but there is a residual multi-year curvilinear trend because we have accounted for a linear trend and the trend in the data is not exactly linear. However, we can now repeat the previous analysis using the detrended time-series, repeating the steps above, resulting in a second plot of an FFT spectrum for these data as shown in Figure 5.10 (which keeps the same vertical-axis limits as Figure 5.8 for comparison purposes).

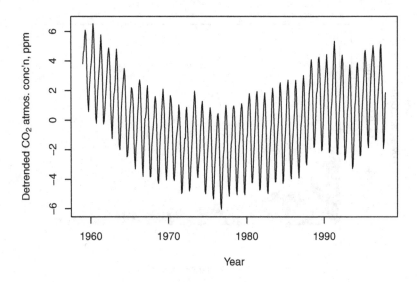

Figure 5.9 Detrended co2 time-series.

```
co2lf <- fft(co2.lin)
co2lop <- data.frame(f.ind, co2lf)
co2lop$f.cal <- co2lop$f.ind * 1/39
co2lop$T <- 39 / co2lop$f.ind
co2lop$mag <- abs(co2lop$co2lf)
co2lop$pwr <- co2lop$mag^2
co2lop.t <- co2lop[1:235,]
co2lop.t$mag[2:234] <- 2 * co2lop.t$mag[2:234]
co2lop.t$pwr[2:234] <- 2 * co2lop.t$pwr[2:234]
plot(co2lop.t$f.cal, co2lop.t$pwr, type="h", log="y",
ylim=c(1e3,1e11), xlab="Frequency, cycles/y",
ylab="Power, arb. units")
```

Figure 5.10 shows evidence of annual cycles, with some evidence of second, third and fourth harmonics, but without most of the noise and noise-like features present in Figure 5.8 because we have largely removed the trend. Also note that the magnitudes of the annual cycle and second harmonic in the detrended data are very similar to those in the raw data, a previously masked fourth is apparent in the detrended data, but the magnitude of the the third harmonic is reduced in the detrended data.

There are many ways of detrending time-series and other data but all endeavour to do the same basic thing: *i.e.* remove the trend and leave only the

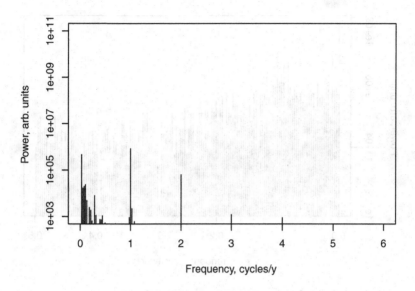

Figure 5.10 Power spectrum, detrended CO_2 time-series.

features of interest around a zero or other constant level. Two other possible detrending options are explored in the Exercises.

5.5 Time-Series with Noisy Spectra

In the preceding examples, we have considered time-series with essentially 'clean' spectra, *i.e.* time-series which are relatively free of complicating features and so the FFT produces spectra with clear markers at the frequencies of the periodic content. Such clean time-series are relatively rare and, more generally, the time-series we investigate will be 'noisy', *i.e.* time-series with features other than the periodic feature(s) of interest and which obscure the features of interest in the FFT spectrum. As an accessible example, we will consider sunspot.year dataset which we met in Chapter 2, as shown in Figure 2.4, and regard that previous analysis as an appropriate initial investigation for the current analysis.

5.5.1 Power Spectrum

We will follow the same procedure as in the previous sections to produce a power spectrum, using the following commands. This is a 289-year time-series

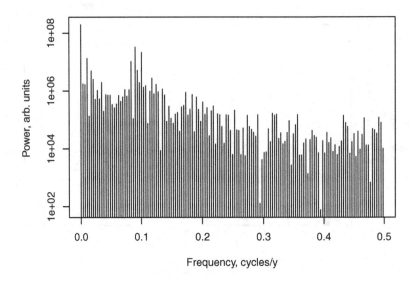

Figure 5.11 Power spectrum, sunspot-year time-series.

(and the unit of time is the year) and so we are considering an N-odd time-series, not N-even as in the previous examples. Thus, the commands for generating the frequency index and aggregating the powers over positive and negative frequencies are different to the previous examples, in accordance with our observations regarding frequency indexing in Chapter 4. The final command below produces a plot as shown in Figure 5.11.

```
f.ind <- c(seq(0,(289-1)/2,1), seq(-((289-1)/2),-1,1))
spf <- fft(sunspot.year)
spop <- data.frame(f.ind, spf)
spop$f.cal <- spop$f.ind * 1/289
spop$T <- 289 / spop$f.ind
spop$mag <- abs(spop$spf)
spop$pwr <- spop$mag^2
spop.t <- spop[1:145,]
spop.t$mag[2:145] <- 2 * spop.t$mag[2:145]
spop.t$pwr[2:145] <- 2 * spop.t$pwr[2:145]
plot(spop.t$f.cal, spop.t$pwr, type="h", log="y",
xlab="Frequency, cycles/y", ylab="Power, arb. units")
```

The spectrum in Figure 5.11 is visibly different to the previous examples, containing 'noise' across the frequency range which obscures the information of interest around the frequencies 0.09–0.10 cycles/year.

The largest non-zero-frequency marker is at $f_{26} = 0.08997$cycles/year, *i.e.* an 11.116-year cycle. This coefficient is *ca.* 40 times larger than the mean of the non-zero-frequency coefficients but is visibly less distinct than the markers in previous examples due to a variety of factors including noise, which we consider in Chapter 8, and the fact that the frequency of the sunspot cycle varies over time, *i.e.* is non-stationary, which we consider in Chapter 7.

The two other large markers in this region of the spectrum are $f_{24} = 0.08304$ and $f_{29} = 0.10035$ cycles/year (*i.e.* 12.042 and 9.966 year cycles respectively). There is also a visibly large marker at $f_3 = 0.01038$ (96.333 years), which supports the observation regarding longer-period cycles we made from the autocorrelogram in Chapter 2, although indicating a ninth sub-harmonic cycle in this analysis rather than the tenth sub-harmonic indicated by the autocorrelation.

5.6 Summary

In this chapter, we have started to characterise what we can typically use the FFT for in the context of time-series analysis. We have also established a basic 'fail-safe' approach for calibrating the FFT harmonic frequencies, generating amplitude and power spectra, and combining the negative-frequency data with the positive-frequency data in order to correctly reveal the frequencies present in a time-series. However, the FFT has limitations, and we will start to explore those in the following chapter.

Exercises

1. In Section 5.2.2.1, if you want to check every coefficient ratio explicitly, repeat the fourth command but without the summary() command. This will list all 240 individual ratios to the console.
2. In Section 5.2.2.2, repeat the first/second/third commands for other pairs of coefficients, according to indexing k and $N - k$ and interpreting these correctly in terms of array indices to select correctly from nottemf, confirming that these are complex-conjugate pairs.
3. Convert the nottem time-series to degrees Celsius and repeat the steps of Section 5.2.3, confirming that the relative magnitudes of the non-zero-frequency coefficients are the same for the two temperature scales but the relative magnitudes of the zero-frequency X_0 coefficient to

the non-zero-frequency coefficients are different for the two temperature scales.

4. Convert the nottem time-series to degrees Celsius and repeat the steps of Section 5.2.4, confirming the ratios of the magnitudes of the powers at 0, 1, 2 cycles/year: the magnitude of the marker at 0 cycles per year is *ca.* 4.3 times the magnitude of the marker at 1 cycle/year; the magnitude of the marker at 1 cycle/year is *ca.* 59 times the magnitude of the marker at 2 cycles/year; and the magnitude of the marker at 2 cycles/year is *ca.* 25 times the average magnitude of the other markers.

5. Repeat the nottem analysis on the tempdub time-series in the TSA library package.

6. Repeat the Hillarys analysis on the PortKembla time-series in the TideHarmonics library package.

7. Repeat the Hillarys analysis on the first two years of the Esperance time-series in the TideHarmonics library package. This time-series contains *ca.* three-week period of missing values (coded NA) in the final year (see Chapter 6 for a consideration of missing data).

8. Repeat the detrending of the co2 time-series using a second-order (quadratic) model. This can be produced in R using the lm() command as follows.

```
co2qm <- lm(co2 ~ poly(tb, 2, raw=TRUE))
```

Then follow the steps in Section 5.4.2, replacing co2lm with co2qm to plot the annual-mean detrended data, FFT the detrended data and plot the FFT output spectrum. Compare the detrended time-series plot to Figure 5.9, and confirm that residual multi-year curvilinear trend is further reduced, and compare the FFT output spectrum plot to Figure 5.10, confirming a further reduction in low-frequency noise and noise-like features.

9. Repeat the detrending of the co2 time-series by subtracting the annual mean values year-wise from the monthly values. This can be done in R using the following commands. The first command creates an empty vector of the correct length, the second a start-of-year (January) index, and the third loops year-by-year, subtracting the annual mean year-wise from the individual monthly values and writing values to the empty vector.

```
co2.dt <- rep(NA, 468)
a.ind <- seq(1, 468, 12)
for(i in a.ind){co2.dt[i:(i+11)] <- co2[i:(i+11)] -
mean(co2[i:(i+11)])}
```

Then follow the steps in Section 5.4.2, replacing co2.lin with co2.dt to plot the annual-mean detrended data, FFT the detrended data and plot the FFT output spectrum. Compare the detrended time-series plot to Figure 5.9, and confirm that the residual multi-year curvilinear trend is further reduced, and compare the FFT output spectrum plot to Figure 5.10, confirming a further reduction in low-frequency noise and noise-like features.

10. Repeat the analysis of Section 5.4 for the co2 time-series in the TSA library package.

Note that the frequency compositions of the different tidal time-series in the TideHarmonics library package are not identical.

6

Constraints on the FFT

We have already encountered some of the limitations of the FFT, *e.g.* specific minimum, maximum and intermediate frequencies in terms of the duration, T, and number of samples, N, of the time-series under investigation. However, we have not looked at how these can affect the FFT spectrum and how we interpret FFT spectra obtained from time-series which contain features the FFT cannot precisely resolve. This chapter considers such issues and the limitations of the FFT.

6.1 Minimum Resolvable Frequency and Low-Frequency Features

If we have a time-series of duration T sampled at N equal intervals, then we know from the basic formalisation of the Fourier series (which underpins the FFT), that the lowest non-zero frequency that the FFT can resolve is $f_1 = 1/T$. However, what happens if we have a time-series which contains features with lower frequencies than $f_1 = 1/T$?

The exact answer for any given time-series depends on the exact nature of the low-frequency features but, in general, the FFT output spectrum will contain low-frequency noise-like features. This arises because, in effect, the FFT synthesises the sub-f_1 features using (predominantly) the low-frequency sinusoids which it can resolve.

6.1.1 Illustration: Low and Zero Frequency Features

To illustrate this, consider the three time-series (or signals) as shown in Figure 6.1, *i.e.* a constant level A (long-dashed line), a linear trend increasing

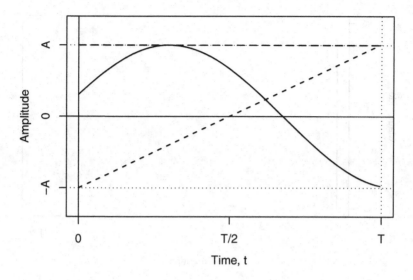

Figure 6.1 Three no/low frequency time-series features.

from $-A$ to A (short-dashed line) and part of a sinusoid with sub-f_1 frequency, in this example $f_{low} = 2/3T$ (solid line).

The FFT amplitude spectrum for the constant level is as we would expect from the formalisation of the FFT, *i.e.* there are no varying features therefore only the f_0 coefficient is non-zero with normalised magnitude $|X_0|/N = A$. This is shown in Figure 6.2.

The FFT amplitude spectrum for the linear trend is as we would expect from the Mauna Loa CO_2 example in Section 5.4 where we observed that if we detrended the time-series, the low-frequency content in the FFT spectrum was reduced. This is shown in Figure 6.3 which reveals that the f_0 coefficient is zero (the mean of this particular line segment is zero), followed by the total f_1 marker with normalised magnitude $|X_1|/N + |X_{-1}|/N = 0.64A$, followed by markers of decreasing magnitude as the frequency increases.

The FFT amplitude spectrum for the part-sinusoid is as we might expect from the previous two cases: we have an overall variation over T which is resolved by the FFT as part constant-level (mean) and part trend. This is shown in Figure 6.4 which reveals the f_0 marker corresponding to the mean level ($|X_0|/N = 0.28A$), followed by the total f_1 marker with magnitude at approximately $|X_1|/N + |X_{-1}|/N = 0.86A$, followed by markers of decreasing magnitude as the frequency increases (more rapidly than for the linear-trend).

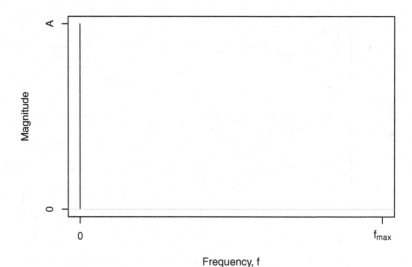

Figure 6.2 FFT spectrum of constant-level time-series.

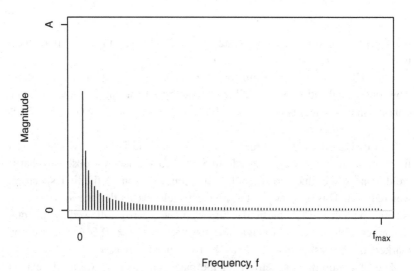

Figure 6.3 FFT spectrum of linear-trend time-series.

6.1.2 Windowing (Tapering)

This refers to multiplying the time-series by a window-function (taper) that reduces the magnitudes of the data progressively to zero as we approach the end-points (*i.e.* $t = 0$, $t = T$) from the centre. Recall from Section 4.6

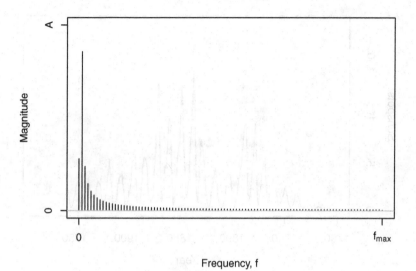

Figure 6.4 FFT spectrum of part-sinusoid time-series.

that because the FFT time-indexing is 2π-periodic, *i.e.* modulo-N, the FFT effectively 'sees' a finite time-series as one periodic sample from an infinite time-series of end-to-end identical samples. Consequently, the FFT 'sees' any end-point discontinuities as low-frequency periodic features. Therefore, windowing (or tapering) removes end-point discontinuities and eliminates the corresponding low-frequency features from the FFT spectrum.

However, the FFT cannot distinguish low-frequencies arising due to co-incidental end-point discontinuities from those arising due to actual end-point features in the data, necessitating an appropriate degree of caution when windowing or tapering datasets. If in doubt, investigate windowed and unwindowed versions of a time-series under investigation.

Commonly used window functions include the Hann function, named after J F von Hann (1839–1921), *i.e.*

$$W(n) = \frac{1}{2}\left(1 - \cos\left(\frac{2\pi n}{N-1}\right)\right)$$

and the Hamming function, named after R W Hamming (1915–1998), *i.e.*

$$W(n) = \alpha - \beta \cos\left(\frac{2\pi n}{N-1}\right)$$

where, conventionally, $\alpha = 0.54$ and $\beta = 0.46$.

Both are raised-cosine windows, and the Hamming function is a refinement of the Hann function. The term 'Hanning', which is sometimes encountered, is a conflation of 'Hann' and 'Hamming'.

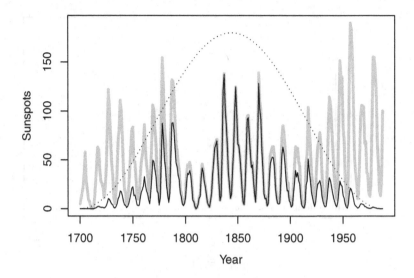

Figure 6.5 Windowed sunspot.year time-series.

The effect of windowing (using a Hann function in this case) is shown schematically in Figure 6.5 for the sunspot.year time-series. In Figure 6.5, the raw data are shown in grey, the Hann-windowed data in black and the dotted curved line shows the shape of the Hann window (not to scale).

6.2 Maximum Resolvable Frequency and Above-Nyquist Features

If we have a time-series of duration T sampled at N equal intervals, *i.e.* a sampling frequency $f_s = N/T$, then we know from the Nyquist–Shannon sampling theorem that the maximum resolvable frequency is $f_{Nyq} = f_s/2$. However, what happens if we have a time-series sampled from a time-function which contains features at higher frequencies than f_{Nyq}?

The complete answer to this question, *i.e.* 'aliasing', arises from the formal derivations of (a) discrete data as sampled from continuous data and (b) and the DFT from the continuous Fourier transform. We did not consider this in Chapter 4, instead choosing an intuitive derivation from Fourier series as being more informative in the current context of investigating periodic features in time-series. However, we can illustrate what happens to frequencies higher than f_{Nyq} by returning to the cyclic nature of the frequency indexing we discussed in Chapter 4.

6.2.1 Aliasing

In Sections 4.6 and 4.7 we observed that the frequency indexing is 2π-periodic, *i.e.* modulo-N, and that this implies that the complete FFT frequency spectrum consists of an infinite number of repeats of the central spectrum indexed on $0 \leq k \leq N - 1$, *i.e.* we have exactly the same spectrum on $N \leq k \leq 2N - 1$, $-N \leq k \leq -1$, and so on to $\pm\infty$.

So far, we have implicitly assumed that all the frequencies in the time-series have not exceeded the Nyquist frequency, f_{Nyq}. The general nature of the FFT spectrum in these circumstances is shown schematically in Figure 6.6 (which is a modified version of Figure 4.7), which shows the full FFT spectrum both in terms of the $\pm f_{Nyq}$ baseband spectrum (solid line) and consecutive copies of it (alternating dashed and solid lines) and also the conventional central-spectrum indexing $0 \leq k \leq N - 1$.

However, if we had a time-series in which the frequencies exceed the f_{Nyq} maximum, we would have overlap between neighbouring instances of the basic spectrum. Frequencies below $-f_{Nyq}$ would overlap into the neighbouring spectrum to the left, and appear as frequencies below $+f_{Nyq}$, and frequencies above $+f_{Nyq}$ would overlap into the neighbouring spectrum to the right, and appear as frequencies above $-f_{Nyq}$. The general nature of the FFT spectrum in these circumstances is shown schematically in Figure 6.7 which, like

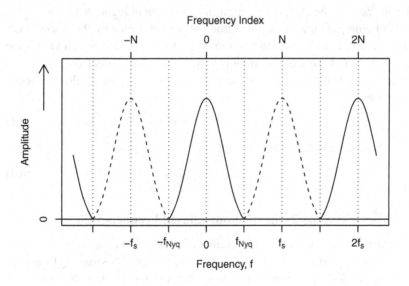

Figure 6.6 Schematic spectrum, all frequencies less than f_{Nyq}.

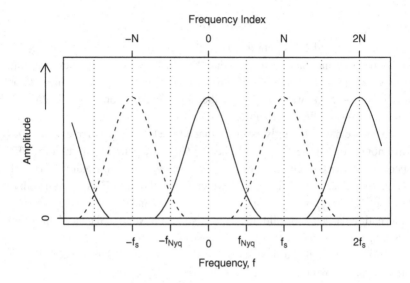

Figure 6.7 Schematic spectrum, frequencies exceeding f_{Nyq}.

Figure 6.6, shows the baseband spectrum (solid line) and consecutive copies of it (alternating dashed and solid lines).

This gives rise to a phenomenon sometimes referred to as 'foldback', because the frequencies above $|f_{Nyq}|$ in the data are folded-back as frequencies below $|f_{Nyq}|$ in the aliased FFT spectrum, as shown in Figure 6.8.

In Figure 6.8, the overlapping true spectra are shown by the dashed lines and the actual FFT spectrum is shown by the solid line, which shows the combination of the true spectrum plus the folded-back tails from the aliased frequencies. In other words, foldback means that a frequency in the data above f_{Nyq}, $e.g.$

$$f_{\text{data}} = f_{Nyq} + \delta f \qquad (6.1)$$

is folded-back in the FFT spectrum as

$$f_{\text{folded}} = f_{Nyq} - \delta f. \qquad (6.2)$$

6.2.2 Illustration: Aliased Frequency

In order to illustrate this, let us consider a hypothetical time-series consisting of a sinusoidal component at frequency $1.1 f_{Nyq}$ sampled at sampling frequency $f_s = 2 f_{Nyq}$. This is shown by the dotted line in Figure 6.9, with the samples indicated by the vertical solid lines at times t_0–t_9. It is clear that there are fewer

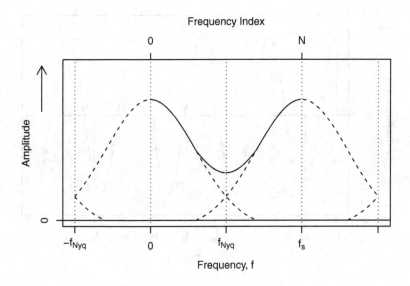

Figure 6.8 Schematic spectrum, aliased.

than two samples per cycle. According to the relationship in Equation 6.1, we have

$$\delta f = f_{\text{data}} - f_{Nyq} = 1.1 f_{Nyq} - f_{Nyq} = 0.1 f_{Nyq}. \qquad (6.3)$$

Therefore, according the relationship in Equation 6.2, we have

$$f_{\text{folded}} = f_{Nyq} - \delta f = f_{Nyq} - 0.1 f_{Nyq} = 0.9 f_{Nyq} \qquad (6.4)$$

and the sinusoid at this frequency is shown by the dashed line in Figure 6.9.

It is clear in Figure 6.9 that the folded frequency ($0.9 f_{Nyq}$, dashed sinusoid) is sampled more than twice per cycle, with samples exactly matching those of the frequency in the data ($1.1 f_{Nyq}$, dotted sinusoid).

6.3 Resolving Intermediate Non-Harmonic Frequencies

If we have a time-series of duration T sampled at N equal intervals, then we know from the basic formalisation of the Fourier series (which underpins the FFT) that the lowest non-zero frequency that the FFT can resolve is $f_1 = 1/T$ and that all other frequencies in the FFT spectrum are harmonics, *i.e.* integer multiples, of this fundamental frequency. However, what happens if we have a time-series which contains features at intermediate, non-harmonic frequencies?

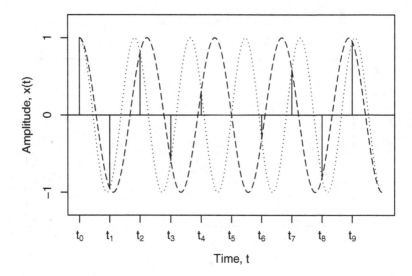

Figure 6.9 Aliased frequency.

The FFT spectrum resulting from an intermediate frequency is as we would expect from the Hillarys tide-height example in Chapter 5 where we observed for the M_2 and O_1 tidal harmonics that instead of single markers at those frequencies, we had 'blurs' of markers at FFT harmonic frequencies centred on those frequencies. This arises from the underlying orthogonality, *i.e.* linear independence, of the sinusoids in the FFT spectrum: sinusoids at intermediate frequencies are not orthogonal to the sinusoids in the FFT spectrum meaning that they are linearly dependent on (correlated with) the sinusoids in the FFT spectrum.

6.3.1 Linear Dependence of Intermediate Frequencies

We can consider the FFT as an orthogonal linear transformation (see Chapter 4) and, thus, we have a mapping of an N-point time-series onto an N-dimensional orthogonal basis, *i.e.* N orthogonal coordinate axes, in which each dimension corresponds to a sinusoid at one of the frequencies in the FFT spectrum. Therefore, a sinusoid with frequency $f_k = k/T$ (*i.e.* the kth FFT harmonic frequency) will have non-zero coefficients (coordinates) on the f_k and f_{-k} axes and zero coefficients on all the other axes, because of the orthogonality of the sinusoidal components in the FFT spectrum. However, a non-harmonic sinusoid with an intermediate frequency, f_i, with. $f_k < f_i < f_{k+1}$ will have non-zero coefficients on all the axes, because this sinusoid is not orthogonal to (and, thus, is linearly dependent on) all the orthogonal harmonic sinusoids those axes represent.

To demonstrate this explicitly, if we rewrite the orthogonality conditions of Section 3.2 in terms of an integer harmonic frequency f_k (resolvable by the FFT) and non-integer frequency f_i (noting $f_k \neq f_i$) we obtain

$$\int_0^T \cos(2\pi f_k t) \cos(2\pi f_i t) dt \neq 0$$

$$\int_0^T \sin(2\pi f_k t) \sin(2\pi f_i t) dt \neq 0$$

$$\int_0^T \sin(2\pi f_k t) \cos(2\pi f_i t) dt \neq 0. \tag{6.5}$$

In practice, this means that given a frequency, f_i, in the data which is intermediate between adjacent harmonic frequencies f_k and f_{k+1} in the FFT spectrum, it will interact most strongly with f_k and f_{k+1}, less strongly with f_{k-1} and f_{k+2}, less strongly again with f_{k-2} and f_{k+3}, and so on. Thus, the spectrum does not contain markers at $f_{\pm i}$ but, instead, contains markers at $f_{\pm k}$ and $f_{\pm(k+1)}$, smaller markers at $f_{\pm(k-1)}$ and $f_{\pm(k+2)}$, smaller markers again at $f_{\pm(k-2)}$ and $f_{\pm(k+3)}$ and so on for all the harmonic frequencies in the FFT spectrum, as shown in Figure 5.5 (and Figure 6.10 in more detail) for the M_2 harmonic in the Hillarys tide-height data. If f_i is much closer to f_k than to f_{k+1} (or vice-versa), then it will yield a much bigger marker at f_k than at f_{k+1} (or vice-versa) and the effective width of the blur in either case will be less than if f_i is approximately equidistant between f_k and f_{k+1}.

How we interpret a spectrum with an intermediate frequency, such as that in Figure 5.5, depends on our knowledge of the data under investigation. Superficially, we have a spectrum which suggests the presence of all the individual frequencies centred on f_i. However, if we know or reasonably expect the frequency f_i to be present in the data with no other frequencies present within a few Δf of f_i, then we can interpret the spectrum as confirming the presence of f_i in the data.

6.3.2 Illustration: Tidal Harmonics

In the Hillarys tide-height example in Section 5.3, the M_2 and O_1 frequency markers are blurred because these frequencies are not integer multiples of the fundamental frequency. In this case, the M_2 and O_1 tidal harmonic frequencies are well known and so interpreting the spectrum is straightforward as the blurs are clearly centred on these frequencies.

However, if we wanted further confirmation, we could make dummy time-series of the same length comprising sinusoids at the M_2 and O_1 tidal harmonic frequencies, and then FFT those dummy time-series and compare their spectra to the Hillarys spectrum. If we have correctly identified the frequencies then, allowing for the presence of other features such as noise in the actual Hillarys data, the M_2 and O_1 blurs in the dummy spectra will have the same shapes across the same frequency intervals as the M_2 and O_1 blurs in the actual Hillarys spectrum. More precisely, allowing for the presence of noise and other features in the Hillarys data, there should be a constant ratio between the amplitudes of the frequency markers in the Hillarys spectrum around the M_2 frequency and the amplitudes of the corresponding markers in the dummy data, similarly for markers around the O_1 frequency. Conversely, if the blurs have different shapes then it is possible that we have incorrectly identified the frequencies and we would need to investigate further.

6.3.3 Sinusoidal Dummy Data

The basic mathematical formulae for creating a time-series of a sinusoid of known frequency (frequency f, period T) and length N are as stated in Section 2.5. More explicitly for the current purpose, we are only interested in frequency, not phase, so we can dispense with the sine (or cosine) term and we do not necessarily need to scale the dummy data although this is sometimes useful, as in the following example.

Thus, for an N-sample time-series of duration T, we need to generate an N-sample sinusoid at the required frequency over T. We can approach this in two ways, either 'unit-less' in terms of the required number of cycles over N intervals indexed $0, 1, 2, \ldots, N - 1$ or 'with-units' in terms of the frequency in cycles per unit time, over times $0, \Delta t, 2\Delta t, \ldots, (N - 1)\Delta t$ for $\Delta t = T/N$ measured in the same units of time.

Taking the M_2 frequency in the Hillarys tide-height data as an example, we can proceed as follows.

According to the unit-less approach, we need to calculate the number of M_2 cycles in 26304 hours, then generate the 26304-interval index, then generate the sinusoid, as follows (using $M_2 = 12.4206012$ to 7 decimal places).

```
Nc <- 26304/12.4206012

ti <- 0:26303
m2 <- cos(2*pi*Nc*ti/26304)
```

According to the with-units approach, it is simplest to work in terms of hours (as recorded) and we need to calculate the frequency (in cycles/hour), then generate the 26304-hour time-base (hourly samples), then generate the sinusoid, as follows.

```
fm2 <- 1/12.4206012

tb <- 0:26303
m2 <- cos(2*pi*fm2*tb)
```

Note that if we decided to work in time-units of days (rather than hours) and frequency in cycles/day, then the with-units approach would be as follows.

```
fm2 <- 24/12.4206012

tb <- (0:26303)/24
m2 <- cos(2*pi*fm2*tb)
```

Whichever route we follow, and provided we are consistent and do not mix units, we obtain the same dummy sinusoid. In practice, depending on the data, some circumstances might favour working in terms of numbers of cycles over intervals indexed $0, 1, 2, \ldots, N - 1$ whereas others might favour working in terms of cycles of given frequency over times $0, \Delta t, 2\Delta t, \ldots, (N - 1)\,\Delta t$.

If we wish to scale the dummy data, *e.g.* for visual comparison as in Figure 6.10, we can use the normalised magnitudes of frequency markers in the real-data spectrum as scale-factors. In this case, we use the normalised magnitude of the biggest of the M_2 markers in the Hillarys spectrum hfop.t ($|X_{2118}| = 1283.031$, noting $26304/2118 = 12.41926$ as in Table 5.1), as follows.

```
Am <- hfop.t$mag[hfop.t$f.ind==2118]/26304
m2 <- Am*m2
```

We can now FFT the scaled dummy m2 time-series, as we did the Hillarys time-series in Section 5.3, as follows.

```
mf <- fft(m2)
m2op <- data.frame(f.ind, mf)
m2op$f.cal <- m2op$f.ind * 1/26304
m2op$T <- 26304 / m2op$f.ind
m2op$mag <- abs(m2op$mf)
m2op$pwr <- m2op$mag^2
m2op.t <- m2op[1:13153,]
m2op.t$mag[2:13152] <- 2 * m2op.t$mag[2:13152]
m2op.t$pwr[2:13152] <- 2 * m2op.t$pwr[2:13152]
```

We can now select an appropriate frequency interval centred on the M_2 frequency ($f_{M2} = 1/12.4206012 = 0.0805114$), *e.g.* $0.080 \leq f \leq 0.081$, extract

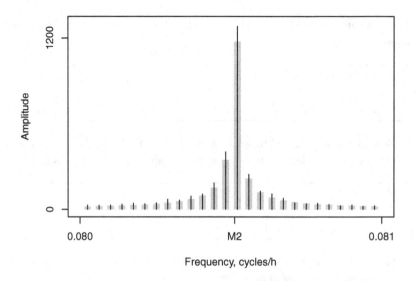

Figure 6.10 M_2 tidal harmonic, Hillarys and dummy-data spectra.

the frequency ranges from both spectra to plot and produce the plot as shown
in Figure 6.10, as follows.

```
m2op.tm <- m2op.t[m2op.t$f.cal>=0.080 &
m2op.t$f.cal<=0.081,]
hfop.tm <- hfop.t[hfop.t$f.cal>=0.080 &
hfop.t$f.cal<=0.081,]
plot(m2op.tm$f.cal, m2op.tm$mag, "h",
xlim=c(0.080,0.081), ylim=c(0,1300), col="grey", lwd=7,
lend=1, xaxt="n", yaxt="n", xlab="Frequency, cycles/h",
ylab="Amplitude")
axis(1, at=c(0.080, 1/12.4206012, 0.081),
labels=c("0.080", "M2", "0.081"))
axis(2, at=c(0, 1200), labels=c("0", "1200"))
par(new=TRUE)
plot(hfop.tm$f.cal, hfop.tm$mag, "h",
xlim=c(0.080,0.081), ylim=c(0,1300), xaxt="n", yaxt="n",
xlab="", ylab="")
```

In Figure 6.10, the broad grey vertical bars show the FFT frequency markers
for the dummy m2 time-series and the narrow black vertical bars show the
FFT frequency markers for the Hillarys time-series. Figure 6.10 shows that
the relationship between the markers in the Hillarys spectrum (hfop.t) and
their counterparts in the dummy spectrum (m2op.t) is highly consistent,
providing confirmatory evidence that the M_2 tidal harmonic is present in the
Hillarys data.

As well as allowing us to construct a dummy sinusoid at an intermediate frequency, it is sometimes useful to construct a dummy sinusoid at one or more FFT harmonic frequencies to check the frequency calibration.

6.4 Re-Windowing Data, Padding

As well as using dummy time-series of appropriate frequencies to help interpret intermediate frequencies in the real data, as in the preceding section, it is sometimes possible and convenient to either truncate or extend ('pad') the time-series under investigation such that an intermediate frequency becomes an FFT harmonic (or very close to an FFT harmonic) of the new different-length time-series. The process of extending a time-series is sometimes referred to as 'zero-padding', although padding with zeros to a required length is strictly only applicable for a time-series with zero mean. When the time-series has a non-zero mean, either subtract the mean value to produce a time-series with zero mean or pad with the mean value, for example if you need to preserve the value of X_0.

As a starting point, if you are seeking to truncate or extend a time-series, in general terms you will have a time-series of duration T (and $T = N\Delta t$) and an intermediate frequency f_i with period $T_i = 1/f_i$ and the ratio T/T_i is not an integer. If subtracting or adding a number of samples, N_d, with $N_d << N$ and $N' = N \pm N_d$, results in a duration $T' = N'\Delta t$ with ratio T'/T_i an integer, then truncating or extending the time-series by this number of samples should not significantly alter the underlying frequency content whilst revealing f_i as a single frequency marker in the FFT spectrum. In practice a number of samples, N', which approximates T'/T_i being an integer will still help to clarify the presence of an intermediate frequency in the data by reducing the effective width of the blur of frequency markers.

Historically, when computers were slower, zero-padding was commonly employed to extend the length of a time-series (signal) to the nearest integer power-of-2 above the length of the time-series in order to use the most optimised FFT algorithm. This is less of a necessity with faster computers (and improved FFT algorithms) but you might still encounter references to this.

6.4.1 Illustration: Tidal Harmonics

We can illustrate this by considering the M_2 tidal harmonic in the Hillarys data, for comparison with the previous example.

The Hillarys data are hourly sampled, therefore we are ideally looking for a number of hours which is an integer multiple of the M_2 period. A little arithmetic shows that 2894 hours is 232.999994 (to 6 decimal places) M_2

tidal cycles, *i.e.* 0.000003% different from 233. Therefore, we would expect that truncating or padding the time-series to an integer multiple of 2894 hourly samples will reveal the M_2 frequency more clearly than the blurs in the preceding spectra. A little more arithmetic indicates that $26304/2894 = 9.0892$, indicating that we could either truncate the Hillarys data to $9 \times 2894 = 26046$ or pad to $10 \times 2894 = 28940$. To illustrate this, we will truncate the data, reset the mean of the truncated time-series to zero, and FFT as follows.

```
hym <- hy[1:26046]
hym <- hym - mean(hym)
hmf <- fft(hym)
f.ind <- c(seq(0,26046/2,1), seq(-(26046/2-1),-1,1))
hmfop <- data.frame(f.ind, hmf)
hmfop$f.cal <- hmfop$f.ind * 1/26046
hmfop$T <- 26046 / hmfop$f.ind
hmfop$mag <- abs(hmfop$hmf)
hmfop$pwr <- hmfop$mag^2
hmfop.t <- hmfop[1:13024,]
hmfop.t$mag[2:13023] <- 2 * hmfop.t$mag[2:13023]
hmfop.t$pwr[2:13023] <- 2 * hmfop.t$pwr[2:13023]
```

We now have both the original and truncated amplitude spectra available. We can normalise these spectra (to remove the scale differences that would arise because $N = 26304$ for the original time-series and $N' = 26046$ for the truncated time-series), extract the frequency ranges from both spectra to plot (using the same interval as in the preceding example) and then produce the plot as shown in Figure 6.11, as follows.

```
hfop.t$magn <- hfop.t$mag/26304
hmfop.t$magn <- hmfop.t$mag/26046
hfop.tm <- hfop.t[hfop.t$f.cal>=0.080 &
hfop.t$f.cal<=0.081,]
hmfop.tm <- hmfop.t[hmfop.t$f.cal>=0.080 &
hmfop.t$f.cal<=0.081,]
plot(hfop.tm$f.cal, hfop.tm$magn, "h",
xlim=c(0.080,0.081), ylim=c(0,0.06), xaxt="n", yaxt="n",
xlab="Frequency, cycles/h", ylab="Amplitude (norm'd)")
points(hfop.tm$f.cal, hfop.tm$magn, pch=45)
axis(1, at=c(0.080, 1/12.4206012, 0.081),
labels=c("0.080", "M2", "0.081"))
axis(2, at=c(0, 0.05), labels=c("0", "0.05"))
par(new=TRUE)
plot(hmfop.tm$f.cal, hmfop.tm$magn, "h",
xlim=c(0.080,0.081), ylim=c(0,0.06), xaxt="n", yaxt="n",
xlab="", ylab="")
```

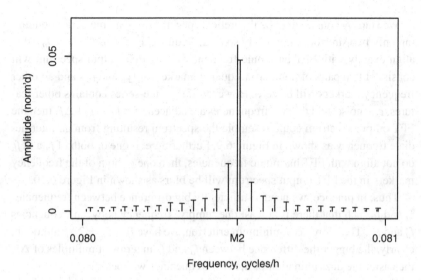

Figure 6.11 M_2 tidal harmonic, original and truncated Hillarys spectra.

In Figure 6.11, the capped vertical bars show the FFT frequency markers for the original Hillarys data (hfop.t) and the uncapped vertical bars show the FFT frequency markers for the truncated Hillarys data (hmfop.t). Note that the frequencies in the two FFT spectra do not align because the two time-series have different lengths (*i.e.* $N' \neq N$, hence different values of Δf). Figure 6.11 shows that the M_2 frequency is very clearly revealed in the truncated spectrum, with almost no blur. This is because we have truncated the length of the time-series to 'tune' the fundamental frequency in the FFT spectrum to produce an FFT harmonic that aligns very closely with the M_2 frequency, *i.e.* f_{2097} noting that $26046/2097 = 12.4206009$ (to 7 decimal places) compared to the value $M_2 = 12.4206012$ that we have been using. This provides confirmatory evidence that the M_2 tidal harmonic is present in the Hillarys data.

6.5 Resolving Adjacent Frequencies

If we have a time-series of duration T with two (closely spaced) frequencies f_a and f_b ($f_b > f_a$), both between f_1 and f_{Nyq}, then this implies there is a maximum value of Δf (corresponding to a minimum value of T) which allows the two frequencies to be resolved distinctly. Thus, if Δf exceeds this maximum value, *i.e.* T is too short, then f_a and f_b will not be resolvable as distinct frequencies.

We start by considering the theoretical special case of a time-series containing only two sinusoids at (closely spaced) frequencies, f_a and f_b ($f_b > f_a$), that align exactly with FFT harmonic frequencies. The FFT output spectrum will consist of two pairs of non-zero frequency markers at $f_{\pm a}$ and $f_{\pm b}$ and all other frequency markers will be zero. However, if the time-series contains other features, e.g. noise, and the two frequencies are adjacent, i.e. $f_b = f_a + \Delta f$, then the FFT output spectrum could resemble the spectrum resulting from an intermediate frequency as shown in Figure 6.9. Furthermore, if one or both of f_a and f_b do not align with FFT harmonic frequencies, then one or both of the frequency markers in the FFT output spectrum will be blurs as shown in Figure 6.10.

Thus, in practice, we need a sufficiently big difference between frequencies f_a and f_b such that there is at least one 'empty' frequency between frequencies f_a and f_b. Therefore, as a minimum criterion, we have $f_b - f_a = 2\Delta f$ although, clearly, the bigger the difference between f_a and f_b in terms of multiples of Δf, the easier the discrimination of the two frequencies will be.

Expressing this with Δf as the subject of the equation, we can obtain expressions for the maximum frequency interval and minimum duration which will allow the FFT to resolve f_a and f_b, i.e.

$$\Delta f_{\max} = \frac{f_b - f_a}{2} \tag{6.6}$$

and, hence

$$T_{\min} = \frac{2}{f_b - f_a}. \tag{6.7}$$

Note that this expression for the minimum duration to resolve two frequencies does not depend on the sampling frequency/interval.

If we wished to have a bigger difference between f_a and f_b in terms of multiples of Δf then we would have, for example for general integer j and interval $j\Delta f$,

$$T = \frac{j}{f_b - f_a}. \tag{6.8}$$

Note that we must use consistent units, i.e. for T in given time-units, f_a, f_b and Δf must be units of cycles per given time-unit, e.g. for T in years, f_a, f_b and Δf must be units of cycles per year.

6.5.1 Illustration: Tidal Harmonics

Time-series containing periodic features often contain closely spaced frequencies, for example the S_2 and M_2 tidal harmonics in the Hillarys data (see Table 5.1).

If we work with the $2\Delta f$ criterion, *i.e.* one empty frequency between the two frequencies of interest, then we can calculate the minimum time-series duration in hours as follows. The two frequencies are $f_{S2} = 1/12 = 0.08333$ and $f_{M2} = 1/12.42 = 0.08051$ cycles/hour and so the minimum duration is

$$T_{\min} = \frac{2}{0.08333 - 0.08051} = 708.74. \tag{6.9}$$

Therefore, a minimum duration 708.74 hours or 29.53 days is required to resolve the S_2 and M_2 tidal harmonics.

The duration of the Hillarys time-series is 26304 hours, which clearly exceeds this criterion and, in fact, $\Delta f = 3.801703 \times 10^{-05}$ cycles/hour for this time-series. This is approximately 1/74 of the difference between f_{S2} and f_{M2} and inspection of the Hillarys FFT spectrum (hfop.t, Section 5.3) reveals that the central markers for S_2 and M_2 tidal harmonics are f_{2192} and f_{2118} respectively, *i.e.* a difference of 74 frequency intervals.

6.6 Missing Data

The FFT cannot transform time-series with missing values. Therefore, if you wish to FFT a time-series with missing values, then you will have to take steps to either work around or replace the missing data, *e.g.* with interpolated or estimated values. Alternatively, provided that you have or can generate a time-index, you could use a periodogram technique, as outlined in Chapter 9.

With regard to the FFT, in the first instance establish where the missing values occur and consider analysing intact sections of the time-series, *i.e.* sections with no missing values separated by (short) sections containing missing values. For example, if the missing values occur at one or both ends, with a main central section of the time-series intact, consider analysing that intact central section. Also, if the missing values are approximately central, consider analysing the two intact approximate half-sections before and after the missing values. Clearly, all such options need to be considered in the context of the length of the intact sections with regard to the required frequency resolution and accounting for the overall duration of the data.

If considering the time-series in smaller intact sections is not an option, *e.g.* because the sections are not long enough or do not adequately cover the overall duration, then a possible option is to replace missing values. Replacing missing values is a process to be undertaken with caution in order to avoid either using an interpolation process which relies on periodic behaviour in the time-series, which could result in the interpolated data producing artefacts in the FFT spectrum, or an estimation process which significantly alters the frequency content, resulting in artefacts in the FFT spectrum. It is not possible to cover every potential approach to replacing missing values but two general criteria are:

- if the time-series under investigation has a small number of isolated, randomly distributed (in time) missing values, then consider using a linear interpolation between the actual values either side of the missing values;
- if the time-series under investigation consists of long intact sections interrupted by a small number of short sections of missing data, then consider replacing the sections of missing data with the mean value of the time-series.

Under any circumstances where you replace missing data, compare the resulting FFT spectrum with FFT spectra from intact sections of the time-series and examine for the presence of frequency markers that are significantly different, reviewing your replacement of missing values accordingly.

6.7 Summary

In this chapter, we have started to characterise some of the limitations of the FFT we might typically expect to encounter in the context of time-series analysis, and the constraints these impose on the time-series we might try to analyse. In particular, it is important to remember that we need a given length of time-series in order to achieve a given frequency resolution and be able to distinguish given closely spaced frequencies. Also, it is important to remember that if confronted with a time-series with missing data, do not replace the missing data by using a technique which makes any use of periodic properties or features in the data.

Exercises

1. Repeat the dummy-data comparison of Section 6.3.3 for the O_1 tidal harmonic, using a period of 25.8193428 hours.
2. Repeat the 'tuning' of Section 6.4.1 for the M_2 tidal harmonic but by padding the time-series to 28940 samples.
3. Repeat the 'tuning' of Section 6.4.1 for the O_1 tidal harmonic. (Hint, try truncating to 24296 samples, approximately 941 O_1 periods.)

4. Repeat the Hillarys analysis on the PortKembla time-series in the TideHarmonics library package.
5. Repeat the Hillarys analysis on the first two years of the Esperance time-series in the TideHarmonics library package. This time-series contains *ca.* 3-week period of missing values (coded NA) in the final year.
6. The Portland dataset in TideHarmonics is similar to Hillarys but has nine consecutive missing values. Identify the nine-hour gap in the data and replace the missing data with the mean value, and then repeat the Hillarys analysis.
7. Using the same approach as for the Hillarys data, analyse the star time-series in the TSA library package and confirm the presence of a large frequency marker at a 24-day cycle and blurred markers centred at 28.57- and 30- day cycles. Create a dummy time-series with a sinusoidal component with a 29-day period and use this to confirm the presence of a *ca.* 29-day cycle in the star time-series.

Note that the frequency compositions of the different tidal time-series in the TideHarmonics library package are not identical.

7

Stationarity and Spectrograms

Up to this point we have implicitly assumed that a time-series has an unchanging frequency composition throughout its duration, *i.e.* broadly, a stationary time-series. However, this is not necessarily the case and some time-series have changing frequency compositions throughout their durations, *i.e.* broadly, non-stationary time-series. For example, the solar irradiance and sunspots data we considered in examples in Chapter 2 and 5 show evidence of non-stationary frequency content. Although for large parts of its duration, as evident in Figure 2.4, there are 'by-eye' regular periodic maxima in that time-series, there is at least one interval, *ca.* 1790–1830 (the Dalton Minimum), where the cycle period is visibly longer (and the maxima smaller) than for the majority of the time-series. Sunspots data are explored in the Exercises.

7.1 Stationarity

Although we are considering stationarity in terms of the relationship between frequency composition and time, there are two widely used formal definitions. With regard to the definitions below, non-stationarity is defined as the opposite of stationarity, *i.e.* a non-stationary process is one where the statistical moments are variant over time.

7.1.1 Strong (Strict) Stationarity

A strongly stationary process is one where all statistical moments of every degree of the process are invariant over time. In other words, a stochastic process where the probability distribution does not vary with time. This is a very strong definition and stronger than we generally need, certainly in the current context. Consequently, unless otherwise specified, we generally work with weak stationarity.

7.1.2 Weak Stationarity

A weakly stationary process is one where the mean (first moment) and variance (second moment) are invariant over time. In other words, a stochastic process where mean and variance do not vary with time. This also means that the autocorrelation (and autocovariance) are invariant over time, *i.e.* the structure of the periodic features is invariant over time. Thus, in the current context, a time-series containing weakly non-stationary features will have a frequency composition that varies over time.

7.2 Short-Time Fourier Transform

There is a variety of techniques for investigating the constancy or inconstancy of the frequency composition of time-series but a basic and informative one is based on the short-time (or short-term) Fourier transform (STFT). In essence, the time-series is divided into sections, generally of equal length at equal intervals through the time-series which can be overlapping or non-overlapping, and the FFT is applied to each section. This yields a time-series of FFT spectra, each of which indicates the frequency composition of its section of the time-series. Therefore, as a whole, this time-series of FFT spectra shows the constancy or inconstancy of the frequency composition through the data time-series. The process of dividing the time-series into sections is often referred to as 'windowing' although this term is used in other related contexts, *e.g.* as in Chapter 6.

In this context, the properties of the STFT are exactly as those of the FFT. This has two important consequences. First, the sampling frequency is the same as for the time-series as a whole and, therefore, the maximum frequency resolvable by the STFT for the sections is the same as for the FFT of the time-series as a whole. Second, the duration of the sections is shorter than that of the time-series as a whole and, therefore, the minimum non-zero frequency and frequency interval resolvable by the STFT for the sections are larger than for the FFT of the time-series as a whole, *i.e.* the frequency resolution for the sections is lower than for the time-series as a whole.

7.2.1 Time and Frequency Resolutions of the STFT

The frequency resolution for the sections is lower than for the time-series as whole which necessitates that we must use sections of appropriate duration for any required frequency resolution and, furthermore, the higher the required frequency resolution, the longer the duration of the sections. However, the longer the duration of the sections, the lower the time resolution: a longer section means that the frequency information is being aggregated over a longer period of time. Therefore, if we want high frequency resolutions then we must use long sections which means low time resolutions. Conversely, if we want high time resolutions then we must use short sections which means low frequency resolutions. More succinctly:

- the longer the section, the higher the frequency resolution (smaller Δf) but the lower the time-resolution (larger T);
- the shorter the section, the lower the frequency resolution (larger Δf) but the higher the time-resolution (smaller T).

We cannot simultaneously have high resolutions in both time and frequency domains. For any given time-series, therefore, there is a balance to be struck between the time and frequency resolutions, *i.e.* between the length of the window in the time-domain and the frequency interval in the frequency-domain, dependent on the features under investigation. In practice, even if we know the minimum frequency (and frequency interval) that we have to account for in a given case, finding the optimum balance between the time and frequency resolutions can require some experimentation.

7.2.2 Uncertainty Principle

As observed in preceding chapters, in terms of Fourier transforms, a time function can be transformed to a frequency function and a frequency function can be transformed to a time function. In other words, time and frequency constitute a pair of variables that are Fourier transform duals. Such variables are referred to as conjugate variables and there are conjugate-variable pairs other than time and frequency, *e.g.* distance and spatial frequency.

The relationship between the time and frequency resolutions observed in the preceding section is an example of the uncertainty principle, notably formalised by W K Heisenberg (1927) in quantum mechanics, and common to all conjugate-variable pairs. In quantum mechanics, the uncertainty principle relates position and momentum and, in outline, states that the higher the precision with which the momentum is known the lower the precision with which the position is known, and *vice-versa*.

However, more generally, uncertainty principles relate to conjugate-variable pairs and assert that it is not possible to increase the precision (decrease the uncertainty) of one member of a conjugate-variable pair without decreasing the precision (increasing the uncertainty) of the other. For Fourier analysis, we can restate the core of the previous section in corresponding terms for time and frequency: the higher the precision with which the frequency is known the lower the precision with which the time is known, and *vice-versa*.

7.3 Spectrograms

The are several types of spectrogram but all are essentially visual representations of the time–frequency relationships in time-series, or equivalent for other data. In the current context, we have STFT spectra calculated at a sequence of steps through a time-series, and with frequency resolution determined by the duration of the window used to section the time-series. The most commonly used graphical format is a two-dimensional one with the horizontal axis for time and the vertical axis for frequency with amplitude of the frequencies represented by a colour or intensity scale. Variations on this are formats with time along the vertical axis and frequency along the horizontal axis and three-dimensional surface-plot formats with horizontal and vertical axes as above but with the amplitude of the frequencies represented by height perpendicular to the axis-plane.

The figures for the following examples were produced with the simple-spectro code in Appendix B. This code was written specifically for this book: it is straightforward and, although it perhaps lacks some more advanced features, is fit for general spectrogram investigations. The default window and step lengths, *i.e.* default window length of one-tenth of the length of the time-series and default step length of one-tenth the length of the window, were chosen as 'failsafes' so that it is possible to produce a basic spectrogram under many circumstances by simply entering the time-series and leaving everything else at the default settings. However, there are no magical default options that are guaranteed to work in every situation: be prepared to experiment.

7.3.1 Overlapping Sections and Time–Frequency Resolution

In general, we use overlapping sections to produce spectrograms, *i.e.* the interval between the mid-points of consecutive sections is less than the section duration and, in a general sequence of sections, the next section starts before the current section ends. This improves precision in the time-domain in that we

obtain a time-series consisting of more closely spaced FFT spectra, but each
FFT spectrum is subject to the time–frequency resolution constraints discussed
above. It is important to note that overlapping the sections does not alter the
basic time–frequency resolution: in effect, overlapping allows us to better align
the sections with changes in frequency content in the time-series and improve
the precision with which we can detect such changes.

7.3.2 Spectrogram: Stationary Data

For our first example, we will investigate the Hillarys time-series. This
is essentially stationary, consisting of well-characterised periodic variations
at the tidal-harmonic frequencies in Table 5.1, plus some lower-frequency
components as evident in Figure 5.5.

A basic full-spectrum spectrogram, as shown in Figure 7.1, using the default
options (except for axis labels) is produced by the following command.

```
simplespectro(Hillarys$SeaLevel, xlab="Time, 01/01/2012
- 31/12/2014", ylab="Frequency, cycles/h", x0="2012",
xT="2015", Ntint=3, greyscale=TRUE)
```

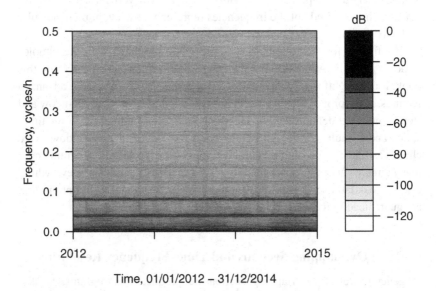

Figure 7.1 Hillarys data, basic spectrogram.

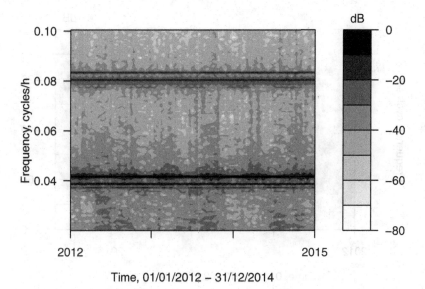

Figure 7.2 Hillarys data, spectrogram detail.

The black/dark-grey horizontal bands show the main frequencies: the band at *ca.* 0.08 cycles/hour corresponds to the semi-diurnal harmonics and the band at *ca.* 0.04 cycles/hour corresponds to the diurnal harmonics. If you want to produce the same spectrogram but with a rainbow colour palette, which will generally make frequency discrimination easier, omit the **greyscale=TRUE** option from the simplespectro() command (and also in the following commands).

A more refined spectrogram to show the semi-diurnal and diurnal tidal harmonics more clearly, as shown in Figure 7.2, is produced by the following command.

```
simplespectro(Hillarys$SeaLevel, Nw=1344, step=240,
fmin=0.02, fmax=0.1, xlab="Time, 01/01/2012 -
31/12/2014", ylab="Frequency, cycles/h", x0="2012",
xT="2015", Ntint=3, greyscale=TRUE)
```

Here, we have used the criteria in Section 6.4 to determine a minimum window length to discriminate the M_2 (12.4206 h) and N_2 (12.6853 h) tidal harmonics, *i.e.* $f_{M2} = 1/12.42 = 0.08051$ and $f_{N2} = 1/12.66 = 0.08333$ cycles/hour:

$$\Delta f = \frac{0.08051 - 0.07900}{2} = 7.559 \times 10^{-4}. \tag{7.1}$$

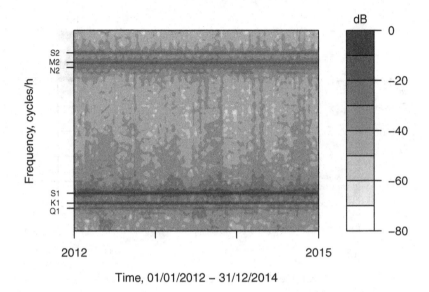

Figure 7.3 Hillarys data, final spectrogram.

hence $T = 1/7.559\times10^{-4} = 1322.88$ hours or 55.12 days. Rounding this to the nearest higher integer number of days gives *i.e.* $T = 1344$ hours (56 days), and we step this at 10-day (240-hour) intervals.

This shows black/dark-grey bands at the S_2, M_2, N_2, S_1, K_1, Q_1 tidal harmonics and, if you wish to confirm this, enter the following commands to recalculate and replot the spectrogram with a calibrated frequency axis, as shown in Figure 7.3 (Figure 7.3 has frequency axis labels slightly offset for clarity).

```
simplespectro(Hillarys$SeaLevel, Nw=1344, step=240,
fmin=0.02, fmax=0.1, xlab="Time, 01/01/2012 -
31/12/2014", ylab="Frequency, cycles/h", x0="2012",
xT="2015", Ntint=3, yextcal=TRUE, greyscale=TRUE)
axis(2, at=c(1/12, 1/12.42, 1/12.66, 1/24, 1/25.82,
1/26.87), labels=c("S2", "M2", "N2", "S1", "K1", "Q1"),
cex.axis=0.7, las=1)
```

7.3.3 Spectrogram: Non-Stationary Data

For our next example, we will investigate the $\delta^{18}O$ record for the presence of Milankovitch (orbital) cycles in the climate record. The frequency content is non-stationary, consisting of different long-period cycles at different times in the Earth's history.

The time-series is stored as the list **JSTR** within the list **OH** in the **RSEIS** package, so we will first extract it as a numerical vector as follows for ease of subsequent operations.

```
o18 <- OH$JSTR[[1]]
```

This is a time-series of 866 samples with constant sample interval of 3 ky (kilo-years), going back in time from most recent to least recent (*i.e.* time getting more negative left to right along the *x*-axis).

We can scope the time-series by producing a time-index (in ky) and plotting as follows.

```
tb <- (0:865)*3
plot(tb, o18, "l", xlab="Time, ky before present",
ylab=expression(paste(delta^"18","O Ratio")))
```

This produces the plot shown in Figure 7.4 which shows evidence of longer-period cycles in the most recent *ca.* 1 million years, with shorter-period cycles further into the past.

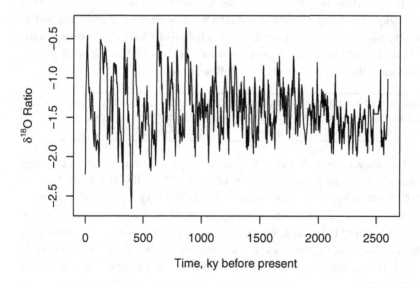

Figure 7.4 The δ^{18}O record.

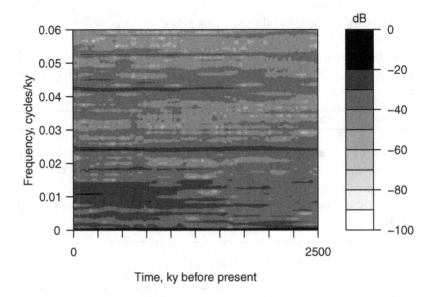

Figure 7.5 Basic $\delta^{18}O$ spectrogram.

The shortest cycle we expect to see has a period of 19 ky (*i.e.* $f = 0.0526$ cycles/ky), which gives us an initial value for the maximum frequency, and the longest cycle we expect to see has a period of 400 ky (*i.e.* $f = 0.0025$ cycles/ky) which suggests that we want a window-length of a few times this to resolve this cycle, hence an initial window length of 1200 ky, *i.e.* 400 intervals. A basic full-spectrum spectrogram, as shown in Figure 7.5, using these options (plus axis labels) is produced by the following command.

```
simplespectro(o18, sput=1/3, fmax=0.06, Nw=400, pstep=1,
xlab="Time, ky before present", ylab="Frequency,
cycles/ky", x0="0", xT="2500", greyscale=TRUE)
```

This shows reasonably clear horizontal stripes (clearer in colour) covering some or all of the period, at values of *ca.* 0.011, 0.018, 0.025, 0.0425 and 0.0525 cycles/ky, *i.e.* periods *ca.* 90, 55, 40, 23, 19 ky respectively, but there is only marginal evidence of a 400 ky cycle (0.0025 cycles/ky) and only in the older part of the record. We can improve the time resolution by shortening the window (noting this will worsen the resolution of any 400 ky cycle) and calibrate the *y*-axis with known Milankovitch/orbital frequencies by suppressing the automatic *y*-axis and adding a separate one calibrated for cycle periods of 400, 95, 72, 55, 41, 23 and 19 ky, as shown in Figure 7.6, as follows.

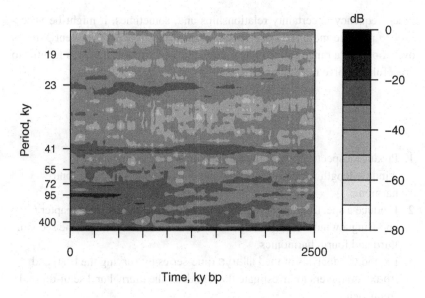

Figure 7.6 Final δ^{18}O spectrogram.

```
simplespectro(o18, sput=1/3, fmax=0.06, Nw=250, pstep=1,
xlab="Time, ky bp", ylab="Period, ky", x0="0",
xT="2500", greyscale=TRUE, yextcal=TRUE)
axis(2, at=c(1/19, 1/23, 1/41, 1/55, 1/72, 1/95, 1/400),
labels=c("19", "23", "41", "55", "72", "95", "400"),
cex.axis=0.9, las=1)
```

Figure 7.6 confirms the presence of cycles at the expected periods for Milankovitch/orbital cycles, *i.e.* cycles observed in data of this type. The resolution of the 95 ky cycle is visibly improved and this confirms that both the 95 and 41 ky cycles are present at the start of the period, but the 95 ky cycle does not persist further back into the record.

7.4 Summary

The spectrogram provides time–frequency information and, noting the uncertainty principle, we gain improvements in time resolution (better time resolution with shorter windows) at the expense of frequency resolution (better frequency resolution with longer windows). Choosing the best window and step lengths can require a little experimentation around the basic underpinning time–frequency uncertainty relationships and, sometimes, it might be necessary to produce more than one spectrogram if the frequency content extends over too great a range for a single combination of window and step lengths to gracefully resolve the required details.

Exercises

1. Produce a spectrogram of the nottem time-series, choosing a window-length which reveals both the annual cycle and the second harmonic.

2. Produce a spectrogram of the detrended co2 time-series (see Chapter 6), choosing a window-length which reveals the annual cycle and the second, third and fourth harmonics.

3. Extend the analysis of the Hillarys time-series by varying the fmin and fmax parameters to investigate the details of the diurnal and semi-diurnal frequencies.

4. Produce spectrograms of the sunspot.year data. Experiment with different combinations of input parameters to show the variation in the length of the *ca.* 11-year sunspot cycles as clearly as possible.

5. Produce spectrograms of the sunspot.month data (also in datasets). Experiment with different combinations of input parameters to show the variation in the length of the *ca.* 11-year sunspot cycles as clearly as possible.

6. Produce spectrograms of the flow time-series in the TSA library package. Experiment with different combinations of input parameters to confirm the presence of a strong 12-month (*i.e.* annual) cycle with a 6-month second harmonic and multi-annual cycles at varying periods through the time-series.

8

Noise in Time-Series

In previous chapters we referred to noise in the general sense of unwanted information which tended to obscure the information of interest but we did not consider noise beyond observing its presence. In this chapter, we will consider noise in a little more detail, enough to be able to interpret a spectrum as is appropriate for an introductory book such as this, but a basic characterisation is as deep as we will go in what is a large and often complicated subject area.

In the context of time-series analysis, a simple definition of noise is unwanted random signals arising from stochastic processes which obscure the periodic features of interest. Noise can arise from a variety of sources including, for example, unknown and unpredictable influences and phenomena in the system from which a time-series is obtained and unwanted variations in the response of the instrumentation used to record a time-series. Whilst it is sometimes possible to configure monitoring systems to reduce noise or process time-series to reduce the presence of noise in subsequent analysis, we need to be clear that neither option is guaranteed and it is all but inevitable that a real time-series will include some form of noise.

There are various categorisations of noise, and various techniques for characterising noise, but an accessible starting point is noise colour. This is a widely used way of describing noise due to its accessibility, if not always its precision.

8.1 Noise Colour

This is an intuitive categorisation of noise that makes an analogy with light. The 'colours' of noise are defined in various mathematical and theoretical ways but all are based on analogies between the power spectra of time-series (signals) and the power spectrum of visible light. We can make the following summary definitions for the colours of noise as generally encountered in, broadly, geoscience time-series. It is important to remember that references to low and high frequency are not absolute and are relative to the phenomena under consideration.

8.1.1 White Noise

In the visible spectrum, in order of increasing frequency in terms of the classical spectral colours, red light has the lowest frequency (with infra-red being lower frequency and not visible to the human eye) through orange, yellow, green, blue, indigo and violet light which has the highest frequency (with ultra-violet being higher frequency and not visible to the human eye). White light comprises all frequencies in the (human) visible spectrum.

By analogy to white light, white noise has a flat power spectrum, *i.e.* uniform power spectral density (*PSD*, power per unit frequency) having no dependence on frequency, which contains all the frequencies in the frequency range equally. Thus, the power spectral density of white noise is invariant with frequency.

8.1.2 Red, Brownian and Pink Noise

Red noise, by analogy with red light, refers to noise in which the PSD decreases with increasing frequency. All 'shades' of red noise can be characterised by the general power–law relationship $PSD \propto f^{\gamma}$ where γ is negative and, in approximate terms, has values generally in the range $-2 \leq \gamma < 0$. Within this general categorisation, shades of redness are sometimes distinguished, *e.g.* Brownian (or Brown) and pink noise. In strict terms, Brownian noise is the noise associated with Brownian motion and is named after R Brown (1773–1858) who first described this form of random-walk motion. It does *not* refer to the colour brown, an unfortunate coincidence which sometimes leads to confusion. Brownian noise is characterised by $PSD \propto f^{-2}$, *i.e.* $\gamma = -2$. At the other end of this range of power–law relationships, pink noise is characterised by $PSD \propto f^{\gamma}$ with, in approximate terms, $-1.5 \leq \gamma \leq -0.5$. Noise with power–law relationship $PSD \propto f^{-1}$, *i.e.* $PSD \propto 1/f$, is sometimes referred to as 'one-over-f' noise.

Variants of red noise are encountered in many geoscience time-series, *e.g.* ocean currents, sea levels, river flows, much climatological data including El Niño phenomena and various temperature time-series, although the actual processes responsible for the noise are not always well understood (*e.g.* Mann and Lees, 1996; Meyers, 2012).

8.1.3 Blue and Violet Noise

Blue noise, by analogy with blue light, refers to noise in which the power spectral density increases with increasing frequency. All shades of blue noise can be characterised by the general power–law relationship $PSD \propto f^\gamma$ where γ is positive and, in approximate terms, has values generally in the range $0 < \gamma \leq 2$. Within this general categorisation, shades of blueness are sometimes distinguished, *e.g.* in approximate terms, blue for lower frequencies with values of $\gamma \approx 1$ and violet for higher frequencies with $\gamma \approx 2$.

Blue noise is much less frequently encountered in geoscience time-series than red noise but, has been sometimes used to describe noise in stratigraphic time-series such as layer thickness in ice-cores (*e.g.* Fisher *et al.*, 1985) and, possibly, applicable to other broadly similar data.

8.2 Statistical Characterisations

8.2.1 Autocorrelation and Autoregression

In Chapter 2 we established that autocorrelation can reveal the presence of periodic features in time-series but now we can develop those ideas to help describe types of noise. If a time-series has a characterisable pattern in a sequence of autocorrelation coefficients (*e.g.* as shown in an autocorrelogram), then it is referred to as an autocorrelated time-series and this implies that values in that time-series have a consistent relationship to previous values in that time-series. We can extend these ideas and, just as we can autocorrelate a time-series against lagged copies of itself, we can also autoregress a time-series against lagged copies of itself to derive an autoregressive (AR) model. Therefore, in autoregressive models the current value of a time-series, x_n, is predicted by a linear combination of previous values of the time-series, *i.e.* $x_{n-1}, x_{n-2}, \ldots$ up to a maximum number of previous terms as appropriate in a given context.

A general autoregressive process of order p, *i.e.* AR(p), means that the current value depends on the previous p values, and has the general form

$$x_n = \alpha + \beta_1 x_{n-1} + \beta_2 x_{n-2} + \cdots + \beta_p x_{n-p} + \varepsilon_n \qquad (8.1)$$

where α is a constant, $\beta_1, \beta_2, \ldots, \beta_p$ are constant coefficients with $\beta_1, \beta_p \neq 0$ and ε_n is a random, white-noise term.

Autoregressive relationships can be used to describe some types of features in time-series as well as some types of noise and, in the context of noise, it is generally first-order relationships, *i.e.* AR(1) models, that are used.

8.2.2 Autoregressive Noise Models

White noise is uncorrelated in the time domain, has a mean value of zero and constant variance. In addition to these properties, Gaussian white noise is white noise that is normally distributed in the time domain. A typical Gaussian white noise time-series is shown in Figure 8.1.

Red (including pink and Brownian) noise has a mean value of zero and constant variance and, like blue noise but unlike white noise, is correlated in the time domain. Red noise can be related to white noise via AR(1) autoregressive relationships of the form

$$x_n = \beta x_{n-1} + \varepsilon_n \sqrt{1 - \beta^2} \tag{8.2}$$

where $\{x_n\}$ is a red noise sequence, $\{\varepsilon_n\}$ is a Gaussian white noise sequence and $\beta = R_1$, the lag-1 autocorrelation coefficient, $0 < \beta < 1$. A typical red noise time-series, for $\beta = 0.8$, derived from the white noise time-series, is shown in Figure 8.2 (using same vertical axis calibration as Figure 8.1 for comparison).

Blue noise has a mean value of zero and constant variance and, like red noise but unlike white noise, is correlated in the time domain. Blue noise can be related to white noise via the same basic AR(1) process as red noise but with negative autocorrelation coefficient, $-1 < \beta < 0$. A typical blue noise

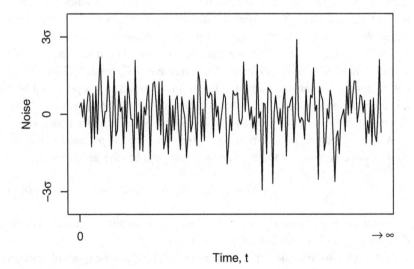

Figure 8.1 Gaussian white noise.

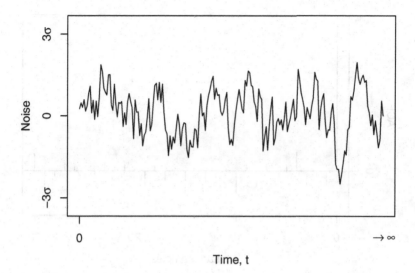

Figure 8.2 Red noise, $\beta = 0.8$.

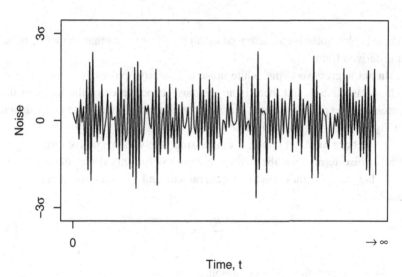

Figure 8.3 Blue noise, $\beta = -0.8$.

time-series, for $\beta = -0.8$ is shown in Figure 8.3 (using same vertical axis calibration as Figure 8.1 for comparison).

Visual comparison of Figures 8.1, 8.2 and 8.3 shows that red noise is characterised by lower frequencies than the white noise it is derived from

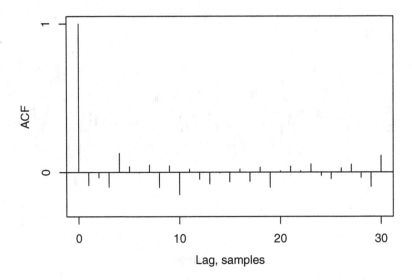

Figure 8.4 White noise autocorrelogram.

whereas blue noise is characterised by higher frequencies than the white noise it is derived from.

In autoregressive terms, white noise is an AR(0) process, *i.e.* the autocorrelation is random and there is no recursive dependence of the terms in the time-series on previous terms. The autocorrelogram for the white noise shown in Figure 8.1 is shown in Figure 8.4.

If we have an AR(1) process comprising solely the recursive term, *i.e.* no white-noise term, then the autocorrelation coefficients decay exponentially with lag, *i.e.* the ratio between general kth and $(k - 1)$th autocorrelation coefficients is

$$\frac{R_k}{R_{k-1}} = \beta \tag{8.3}$$

for $k > 0$ and $R_0 = 1$ by definition.

However, the inclusion of the white-noise term in the red-noise AR(1) process modifies this behaviour. For small lags, the recursive term dominates and the low-index autocorrelation coefficients decay according approximately to Equation 8.3. For larger lags, the white-noise term dominates and the high-index autocorrelation coefficients behave approximately as a white-noise autocorrelation. The transition between these behaviours is characterised by a decay-time, τ,

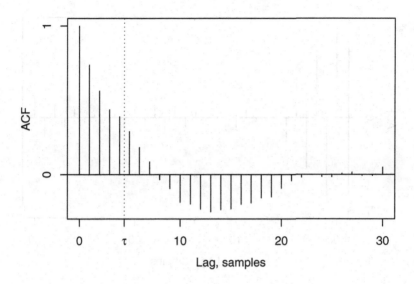

Figure 8.5 Red noise autocorrelation, $R_1 = 0.8$.

$$\tau = -\frac{\Delta t}{\ln(|\beta|)} = -\frac{\Delta t}{\ln(|R_1|)} \tag{8.4}$$

where Δt is the lag-interval (generally, but not always, equal to the sampling interval in the time-series).

The autocorrelogram for the red noise shown in Figure 8.2 is shown in Figure 8.5, with the decay-time shown by a vertical dotted line.

The autocorrelogram for the blue noise shown in Figure 8.3 is shown in Figure 8.6, and the alternating signs of the autocorrelation coefficients is clear.

8.3 Power–Law Relationships

In Section 8.1 we introduced the basic power–law relationship, $PSD \propto f^\gamma$, between power spectral density and frequency which characterises noise colour according the value of γ. If we express the relationship between power spectral density and frequency as an equation with constant of proportionality, α, then we obtain

$$PSD = \frac{P_f}{\Delta f} = \alpha f^\gamma \tag{8.5}$$

where P_f is the power in interval Δf at frequency f.

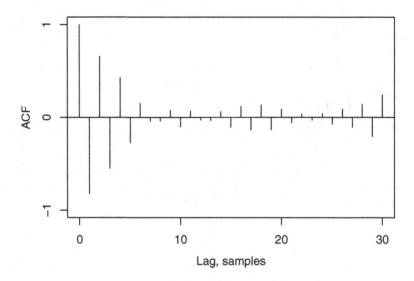

Figure 8.6 Blue noise autocorrelation, $R_1 = 0.8$.

We can rearrange Equation 8.5 to make P_f the subject and take logarithms of both sides to obtain

$$\log\left(P_f\right) = \gamma \log\left(f\right) + \log\left(\alpha \Delta f\right) \tag{8.6}$$

which is a linear relationship between $\log\left(P_f\right)$ and $\log\left(f\right)$ with gradient γ and intercept $\log\left(\alpha \Delta f\right)$.

Note that the gradient of this linear relationship is not affected by the units of power or frequency: any power or frequency scaling factor affects the intercept but not the gradient. Therefore, it does not matter whether we normalise the power, or scale power to variance, or consider frequency in units of harmonic index or cycles per unit-time: the gradient of the logarithmic form of the power–law equation represents the value of the power law. Thus, if we plot the power spectrum using logarithmic axes, any power–law relationships between power spectral density and frequency will be revealed by (approximate) straight-line segments in the plot.

The white noise power spectrum is shown in Figure 8.7. In Figure 8.7 the horizontal dashed line represents the implicit power law with gradient $\gamma = 0$.

The red noise power spectrum is shown in Figure 8.8. In Figure 8.8 the inclined dashed line represents the red-noise power law with gradient $\gamma \approx -1.1$ in this example.

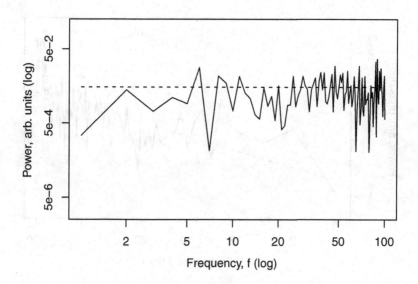

Figure 8.7 White noise power spectrum, logarithmic axes.

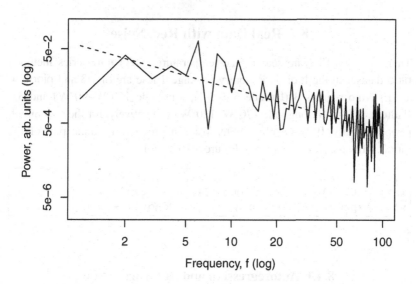

Figure 8.8 Red noise power spectrum, logarithmic axes.

The blue noise power spectrum is shown in Figure 8.9. In Figure 8.9 the inclined dashed line represents the blue-noise power law with gradient $\gamma \approx 1.2$ in this example.

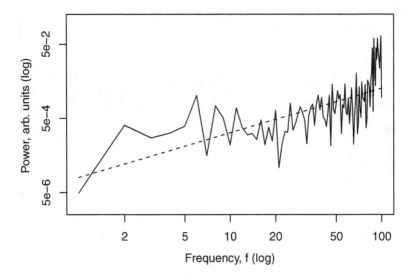

Figure 8.9 Blue noise power spectrum, logarithmic axes.

8.4 Real Data with Red-Noise

There are two El Niño sea surface temperature (SST) time-series included nino dataset in the tseries R library package These are nino3 and nino3.4, for El Niño regions 3 (latitude 5°S–5°N, longitude 150°W–90°W) and 3.4 (latitude 5°S–5°N, longitude 170°W–120°W) respectively, for the 598-month period January 1950–October 2009. We can produce a simple plot of the nino3.4 time-series, as shown in Figure 8.10, as follows.

```
plot(0:597/12, nino3.4, "l", xlim=c(0,599/12),
xaxp=c(0,600/12,10), ylim=c(24,30), xlab="Years",
ylab=expression("Temperature, " * degree * C))
```

8.4.1 Autoregression and Autocorrelation

We can use R's built-in ar() command to estimate the autoregressive constant, β, and other information as follows. We need to include the order.max option to prevent the algorithm finding higher-order terms resulting from the non-red-noise properties of the time-series.

```
nino.ar <- ar(nino3.4, order.max=1)
```

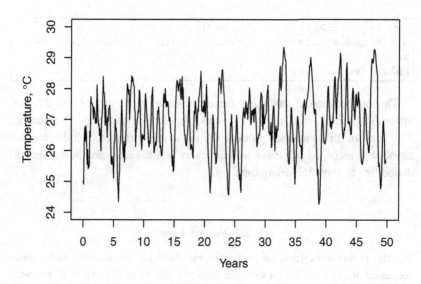

Figure 8.10 The nino3.4 time-series.

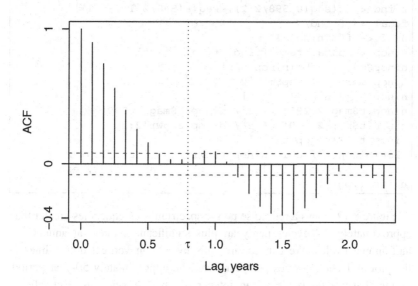

Figure 8.11 The nino3.4 autocorrelogram.

This yields the autoregressive constant $\beta = 0.9$. From there, we can calculate the decay-time, τ, and plot the autocorrelogram shown in Figure 8.11 as follows.

```
tau <- (-1/12)/log(nino.ar$ar)
acf(nino3.4, xlab="Lag, years", main=NA, yaxt="no")
axis(2, at=c(-0.4,0,1), labels=c("-0.4","0","1"))
abline(v=tau, lty=3)
```

The first command calculates the decay-time, τ, in years ($\tau = 9.5$ lag-intervals (months), *i.e.* $\tau = 0.79$ years, for $\Delta t = \frac{1}{12}$ year). The second, third and fourth commands produce the autocorrelogram (which detects that nino3.4 is recorded in years), add a clearer vertical axis, and add a vertical dotted line to show the decay-time.

8.4.2 Power Law

For the power spectrum, as shown in Figure 8.12, we use the same basic sequence of commands that we have used previously in Chapter 5, as follows.

```
T <- 598/12
f.ind <- c(seq(0,598/2,1), seq(-(598/2-1),-1,1))
f.cal <- f.ind/T
ninf <- fft(nino3.4)
ninop <- data.frame(f.ind, f.cal, ninf)
ninop$mag <- abs(ninop$ninf)
ninop$pwr <- ninop$mag^2
ninop.t <- ninop[1:300,]
ninop.t$mag[2:299] <- 2 * ninop.t$mag[2:299]
ninop.t$pwr[2:299] <- 2 * ninop.t$pwr[2:299]
ninop.t <- ninop.t[-1,]
plot(ninop.t$f.cal, ninop.t$pwr, type="l", log="xy",
xlab="Frequency, cycles/y (log)", ylab="Power, arb.
units (log)")
```

Figure 8.12 shows a red-noise power spectrum for frequencies greater than approximately 0.2 cycles per year, plus significant markers at annual and half-annual cycles. We can confirm this by adding vertical dotted lines to the plot at 1 and 2 cycles per year (over the approximately 50-year period, noting that we do not have an integer number of years) respectively, as follows.

```
abline(v=c(1,2), lty=3)
```

To estimate the power law and add a regression line to the plot, we need to regress the logarithm of the power against the logarithm of the frequency, for frequencies greater than 0.2 cycles per year, as follows.

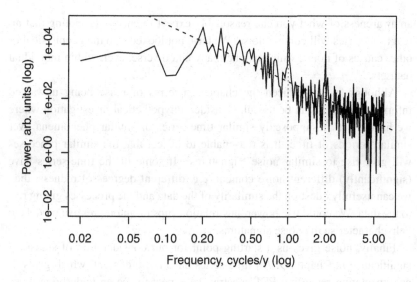

Figure 8.12 The nino3.4 power spectrum, showing power law.

```
ninlog <- log10(ninop.t[,c("f.cal", "pwr")])
ninplm <- lm(pwr ~ f.cal,
data=ninlog[ninlog$f.cal>log10(0.2),])
abline(ninplm, lty=2)
```

The first command extracts the frequency and power to a new data-frame and takes the logarithms and the second regresses the logarithm of the power against the logarithm of the frequency for frequencies greater than 0.2 cycles per year. The third command adds a dashed line to the plot showing the power law.

We can print a regression summary to the console as follows.

```
summary(ninplm)
```

This shows that the gradient, *i.e.* power law $\gamma \approx -2.1$.

In order to estimate the red-noise spectrum more accurately, we should ideally model the overall annual cycle and subtract this from the time-series and analyse the residual time-series.

8.5 Summary

In this chapter we have started to consider noise as can occur in, broadly, geoscience time-series. This is only a brief introduction, enough to give you an awareness of what you can reasonably expect to encounter, noting that all real time-series will contain noise. We have not looked into the geophysical or other causes of noise, which can be many and diverse, even within individual datasets.

Although we introduced the chapter in terms of noise being unwanted information, noise can be useful. Consider a hypothetical investigation where we have a set of supposedly similar time-series of similar phenomena over similar periods of time. It is reasonable to expect that the similar processes will give rise to similar noise 'signatures'. If some of the time-series have (significantly) different noise content, *e.g.* different degrees of redness, then we can usefully question the similarity of the data and the processes giving rise to the data. We could use power–law relationships or autoregressive models to help characterise the noise signatures.

Finally, noise gives us a starting point for our consideration of statistical significance in Chapter 9. All time-series have noise content, which gives us an expectation regarding FFT spectra: we expect to see an underlying noise spectrum whether or not there is deterministic frequency content. We can use the relative magnitudes of deterministic frequency markers compared to background noise markers as a basis for investigating statistical significance.

Exercises

1. Repeat the above analysis (Section 8.4) for the nino3 time-series. How similar are the two noise power laws?
2. Characterise the noise power law (calculate γ) for the treerings time-series, in the datasets package.
3. Characterise the noise power law (calculate γ) for the camp time-series, in the tseries package.
4. Characterise the noise power law (calculate γ) for the $\delta^{18}O$ time-series from OH, in the RSEIS package (see Chapter 7).

9

Periodograms and Significance

Thus far, at least until the end of Chapter 8, we have considered the FFT as a technique for representing a time-series in terms of a spectrum of harmonic frequencies which estimates the true frequency composition. Also, we have established the constraints of the FFT in terms of the frequencies it can and cannot 'see'. However, we have not considered the statistical significance of FFT spectra, *i.e.* the probability that a given spectrum, or markers of given amplitudes at given frequencies in a spectrum, are deterministic or could occur by chance as some form of noise or other stochastic phenomenon in the data under investigation. As a starting point, we will return to Chapter 2 and extend our consideration of correlation and regression techniques because both these are least-squares techniques which provide measures of explained and unexplained variance which enable us to consider variance ratios and statistical effect-size.

Also, we have confined our consideration to equal-interval time-series and not all time-series are equal-interval. Although the FFT is restricted to equal-interval time-series, the correlation and regression techniques considered in Chapter 2 are not and these have developed via separate routes into a range of techniques which can be grouped as Least-Squares Spectral Analysis (LSSA). We can note in passing that, under some circumstances, it might be possible to resample an unequal-interval time-series onto a sequence of equal-interval samples. However we must also note that, as with the replacement of missing data discussed in Chapter 6, resampling needs to be undertaken with caution to avoid introducing artefacts.

In this chapter, before returning to a consideration of explained and unexplained variance in FFT spectra, we will consider two periodogram approaches for analysing periodic content in unequal-interval time-series and, under some circumstances equal-interval time-series with missing data, which also provide measures of statistical significance (p-value). A "periodogram" is an estimate of the power spectrum of a time-series (signal), and the term was introduced by F A F Schuster (1897 and 1898) where he introduced the periodogram approach. In his 1897 paper, Schuster was investigating unequal-interval time-series of earthquake incidence: in essence, the time-series that he considered and developed his periodogram technique for were sequences of events, with implied equal magnitudes (effectively unit magnitudes), at unequal intervals. This type of time-series is distinctly different to the time-series that we have considered for the FFT, which are sequences of samples of varying magnitudes at equal intervals.

Schuster's approach has been developed and its links with Fourier and correlation and regression techniques have been explored by many people chiefly, in the immediate context, by J D Scargle (1982) who developed earlier work by N R Lomb (1976). Their papers took forward work by P Vaníček (1969, 1971).

9.1 The Schuster, or Classical, Periodogram

Schuster developed his approach to investigate the possible presence of periodic behaviour in the incidence of earthquakes. In this approach, the time of each event is converted from the standard units used to record the data to a unit based on the cycle period under investigation, T_k, then mapped onto phase around the cycle according to that period, *i.e.* equating T_k to 2π. Each event is considered as a unit-magnitude event represented as a unit-vector with orientation (around a circle) determined according to its phase. This is equivalent to a two-dimensional uniform-step random-walk process, which is the basis for estimating the probability and significance.

For an N-sample time-series of duration T, not necessarily equal-interval, of events $\{x_n\}$ and the times at which they occur $\{t_n\}$, for $n = 0, 1, 2, \ldots, N-1$, we have

$$\{(x_n, t_n)\} = (x_0, t_0), (x_1, t_1), \ldots, (x_{N-1}, t_{N-1}) \tag{9.1}$$

with, in this case, $x_0 = x_1 = x_2 = \cdots = x_{N-1} = 1$.

For a cycle of interest with frequency f_k and period T_k, where f_k is implicitly assumed to be an integer multiple of the fundamental frequency $f_1 = 1/T$, we re-express the times $\{t_n\}$ in terms of multiples of T_k, as shown schematically in Figure 9.1.

From there, we re-express the times (in terms of multiples of T_k) as times modulo-T_k, *i.e.* we discard the integer parts of the times such that $0 \le t_n < T_k$

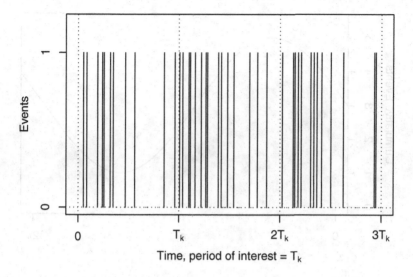

Figure 9.1 Schematic time-series of events at unequal intervals.

for $n = 0, 1, 2, \ldots, N-1$, and convert those times to phases, *i.e.* $\phi_n = t_n \cdot 2\pi / T_k$, to obtain the series

$$\{(x_n, \phi_n)\} = (x_0, \phi_0), (x_1, \phi_1), \ldots, (x_{N-1}, \phi_{N-1}) \qquad (9.2)$$

where $0 \leq \phi_n \leq 2\pi$ for $n = 0, 1, 2, \ldots, N - 1$. From there, we choose an appropriately sized phase interval ('bin') as a fraction of 2π, according to the data, and then produce a histogram as shown in Figure 9.2.

At this point, we can fit a sinusoid of period T_k as indicated by the solid line in Figure 9.2, an approach which underpins many periodogram techniques including Lomb–Scargle, but Schuster himself suggested an alternative approach, as follows, as being more intuitively reasonable.

Instead of fitting a sinusoid, Schuster considered the histogram as a set of vectors in which the magnitudes were given by the numbers of events in the histogram bins and the orientations were given by the bin-central phases around a complete rotation (2π radians). These vectors can then be summed to obtain the resultant vector and the probability of a vector of that magnitude arising by chance can be used to evaluate the significance of the period T_k in the time-series.

Note that Schuster described this approach in terms of the histogrammed data because he did not have access to the computing hardware and software that is now routinely available and so needed to simplify the calculations to those manageable by hand. However, it is not theoretically necessary to

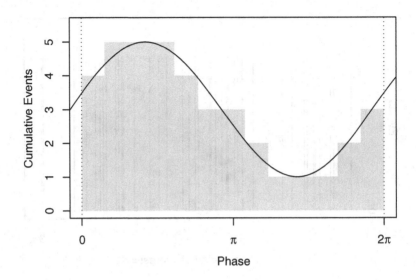

Figure 9.2 Schematic histogram of events according to phase.

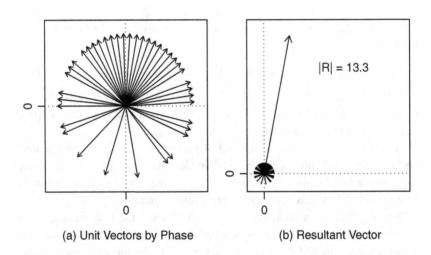

(a) Unit Vectors by Phase   (b) Resultant Vector

Figure 9.3 Vector representation and summation.

aggregate the events into a histogram for this vector-sum approach and, with computer hardware and software tools now available, we can consider the same approach for individual events as shown schematically in Figure 9.3.

Figure 9.3a shows a polar plot of the events shown in Figure 9.1 represented as unit-vectors according to their phase around a cycle of period T_k, and there is a visible clustering of events towards the top of the plot at an angle a little less than $\pi/2$ radians. Figure 9.3b shows the same information as Figure 9.3a plus the resultant vector, as obtained from the calculations outlined in the following sections.

9.1.1 Vector Summation

Having mapped the times $\{t_n\}$ onto phases $\{\phi_n\}$ according to cycle-period T_k, we resolve these coordinates into Cartesian form and then sum to find the resulting vector, r, with Cartesian components (r_X, r_Y) given by

$$r_X = \sum_{n=0}^{N-1} x_n \cos(\phi_n)$$

$$r_Y = \sum_{n=0}^{N-1} x_n \sin(\phi_n) \tag{9.3}$$

with $x_0 = x_1 = x_2 = \cdots = x_{N-1} = 1$ for individual events.

We then resolve back to magnitude and phase (polar) coordinates (r, ϕ), *i.e.*

$$|r|^2 = \sqrt{r_X^2 + r_Y^2}$$

$$\phi = \arctan(r_Y/r_X). \tag{9.4}$$

As well as quantifying the probability that the resultant 'random-walk' distance from the origin could arise by chance, we can, if required, map the phase back onto time to determine when the maximum occurs in the cycle with respect to the time-reference.

The result of the summation is a vector with magnitude $0 \leq |r| \leq N$. At one extreme, if the event-incidence has no dependence on phase then the result of summing the individual event-vectors will be a vector with $|r| \approx E(r)$, the expectation (see below). At the other extreme, if all the events occur at exactly the same phase, then the result will be a vector with $|r| = N$, at some phase around the cycle.

For the schematic example illustrated in the preceding paragraph, these calculations yield:

$$|r| = 13.3 \text{ units}$$

$$\phi = 1.38 \text{ radians.} \tag{9.5}$$

9.1.2 Probability and Significance

If we wish to use this approach as a statistical test, we need null and alternative hypotheses, *i.e.*

- null hypothesis, H_0, the samples occur randomly, have no dependence on phase;
- alternative hypothesis, H_1, the samples occur deterministically, have significant dependence on phase.

The probability density function for an N-event time-series (*i.e.* N-step random-walk) to result in a distance $|r|$ from the origin is given by the Rayleigh distribution, as presented by Lord Rayleigh (1880), *i.e.*

$$\text{PDF}(|r|) = \frac{2\,|r|}{N} e^{-|r|^2/N}. \tag{9.6}$$

The Rayleigh distribution is more usually expressed in terms of a scale parameter, σ, as

$$\text{PDF}(|r|) = \frac{|r|}{\sigma^2} e^{-|r|^2/2\sigma^2} \tag{9.7}$$

with $\sigma = \sqrt{N/2}$ in this case.

This yields a cumulative distribution function (CDF) for distances up to $|r|$ occurring by chance

$$\text{CDF}(|r|) = 1 - e^{-|r|^2/N}. \tag{9.8}$$

Thus, the 'tail' probability of distances exceeding $|r|$ by chance (*cf.* p-value), is

$$p = 1 - \text{CDF}(|r|)$$
$$= e^{-|r|^2/N}. \tag{9.9}$$

Note that the Rayleigh distribution has a non-zero expectation, *i.e.* the expected resultant of a set of randomly oriented vectors is greater than zero. The expectation is

$$E(r) = \frac{1}{2}\sqrt{N\pi} \tag{9.10}$$

and this represents the null-hypothesis outcome.

This is shown in Figure 9.4, for the schematic example illustrated in the preceding paragraph: *i.e.* $N = 39$, $\sigma = 4.42$, $E(r) = 5.53$ and $|r| = 13.3$. For this hypothetical example, $p\,(|r| \geq 13.3) = 1.05\%$, *i.e.* this is a statistical p-value that a vector of this magnitude or greater occurs by chance (upper tail, shown shaded in Figure 9.4).

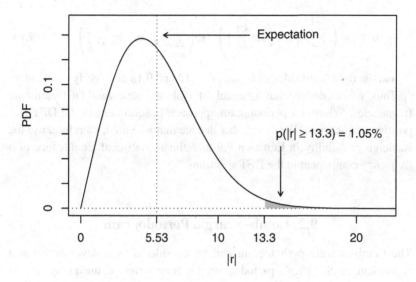

Figure 9.4 Rayleigh distribution for $N = 39$ and $|r| = 13.3$.

9.1.3 Comparing the Schuster Periodogram to the DFT

If we consider the time-series in Equation 9.1 as being equal-interval, then the 'standard' form for the kth positive frequency coefficient in the DFT spectrum is:

$$X_k = \sum_{n=0}^{N-1} x_n e^{-2\pi i \frac{kn}{N}}.$$

$$= \sum_{n=0}^{N-1} x_n \left(\cos \left(2\pi \frac{kn}{N} \right) - i \sin \left(2\pi \frac{kn}{N} \right) \right)$$

$$= \sum_{n=0}^{N-1} x_n \cos \left(2\pi \frac{kn}{N} \right) - i \sum_{n=0}^{N-1} x_n \sin \left(2\pi \frac{kn}{N} \right). \qquad (9.11)$$

From there, we have the power for the kth positive frequency coefficient:

$$|X_k|^2 = \left(\sum_{n=0}^{N-1} x_n \cos \left(2\pi \frac{kn}{N} \right) \right)^2 + \left(\sum_{n=0}^{N-1} x_n \sin \left(2\pi \frac{kn}{N} \right) \right)^2. \qquad (9.12)$$

The Schuster periodogram result for that same frequency, expressing the vector components r_X and r_Y explicitly, is:

$$|r_k|^2 = \left(\sum_{n=0}^{N-1} x_n \cos\left(2\pi \frac{kn}{N}\right)\right)^2 + \left(\sum_{n=0}^{N-1} x_n \sin\left(2\pi \frac{kn}{N}\right)\right)^2 \qquad (9.13)$$

and the right-hand sides of Equations 9.12 and 9.13 are clearly identical.

Thus, we can deduce that for equal-interval time-series, and DFT-harmonic frequencies, Schuster's periodogram approach is equivalent to the DFT (for positive frequencies). We can further deduce that we could, therefore, use the Rayleigh probability in Equation 9.9 to estimate statistical significance of a frequency component in the DFT spectrum.

9.2 Lomb–Scargle Periodogram

The Lomb–Scargle periodogram can be considered as a development and refinement of Schuster's periodogram for time-series of unequally spaced samples rather than unequally spaced individual events. It has the same underlying form as Schuster's periodogram but normalises the resultant random-walk distance differently, *i.e.* for the general time-series indicated in Equation 9.1 and the kth frequency term, we have

$$|r_k|^2 = \frac{1}{2}\left(\frac{\left(\sum_{n=0}^{N-1} x_n \cos\left(2\pi f_k (t_n - \tau_k)\right)\right)^2}{\sum_{n=0}^{N-1} \cos^2\left(2\pi f_k (t_n - \tau_k)\right)}\right.$$
$$\left.+\frac{\left(\sum_{n=0}^{N-1} x_n \sin\left(2\pi f_k (t_n - \tau_k)\right)\right)^2}{\sum_{n=0}^{N-1} \sin^2\left(2\pi f_k (t_n - \tau_k)\right)}\right) \qquad (9.14)$$

where the time offset, τ_k, is given by

$$\tau_k = \frac{1}{4\pi f_k} \arctan\left(\frac{\sum_{n=0}^{N-1} \sin\left(4\pi f_k t_n\right)}{\sum_{n=0}^{N-1} \cos\left(4\pi f_k t_n\right)}\right) \qquad (9.15)$$

and f_k is not necessarily an integer multiple of the fundamental frequency $f_1 = 1/T$.

The time-offset τ_k arises from Lomb's analysis (more specifically, from the diagonalisation of a quadratic-form matrix used in the least-squares formalisation) and ensures that the sinusoids at frequency f_k are orthogonal over the entire duration T (not just over whole cycles at f_k). This is important because it means the sine and cosine terms in Equation 9.14 have zero covariance even for fractional cycles and so can be separated as in the expression.

For frequency f_k (period $T_k = 1/f_k$, if k represents an integer number of whole cycles or half cycles over the duration T, then the sinusoids $\cos(2\pi f_k t_n)$ and $\sin(2\pi f_k t_n)$, for $n = 0, 1, 2, \ldots, N-1$, are orthogonal over T and $\tau_k = 0$. In such cases, both individual denominators in Equation 9.14 are equal to $N/2$ and the expression reduces to Schuster's expression. However, for all other values of k the sinusoids $\cos(2\pi f_k t_n)$ and $\sin(2\pi f_k t_n)$ are not orthogonal over T and $\tau_k \neq 0$ but the sinusoids $\cos(2\pi f_k (t_n - \tau_k))$ and $\sin(2\pi f_k (t_n - \tau_k))$ are orthogonal over T.

9.2.1 Probability and Significance

The derivation of a statistical significance is similar to that for the Schuster periodogram but differs in that it accounts for the probability of the maximum value in a set of J independent frequencies occurring by chance, whereas Schuster's formalisation considers an individual frequency.

Starting with a single frequency, f_k, and normalising the power (which is equivalent to considering unit steps in a two-dimensional random walk, as in Schuster's formalisation) we obtain

$$z = \frac{r_k^2}{\sigma^2} \tag{9.16}$$

where σ^2 is the variance of the time-series.

From there, we have the probability density function for z

$$\text{PDF}(z) = e^{-z} \tag{9.17}$$

and the cumulative distribution function for z is

$$\text{CDF}(z) = 1 - e^{-z}. \tag{9.18}$$

For the maximum value in a set of J independent frequencies, *i.e.* $Z = \max(z_J)$, we have the modified cumulative distribution function

$$\text{CDF}(Z) = \left(1 - e^{-Z}\right)^J \tag{9.19}$$

which yields a 'tail' probability of maximum normalised power exceeding Z by chance in a set of J independent frequencies (*cf.* p-value)

$$p = 1 - \text{CDF}(z)$$
$$= 1 - \left(1 - e^{-Z}\right)^J. \tag{9.20}$$

9.2.2 Effective Nyquist Frequency

Unlike the case of equal-interval time-series, where the Nyquist frequency (Section 4.8) is well defined, the concept of the Nyquist frequency in unequal-interval time-series is considerably less straightforward. In an unequal-interval time-series, there are some regions where the sample intervals are shorter, *i.e.* instantaneous sampling frequencies are higher, and some regions where the sample intervals are longer, *i.e.* instantaneous sampling frequencies are lower. Higher instantaneous sampling frequency implies higher instantaneous Nyquist frequency and, conversely, lower instantaneous sampling frequency implies lower instantaneous Nyquist frequency. Without going into a full analysis, which is beyond the scope of this book, we need to calculate an average sampling frequency in order to estimate the effective Nyquist frequency as the upper limit to the frequencies we can resolve in a time-series.

For a basic criterion, we can work with an initial statement in Appendix D in Scargle's (1982) paper which states that the maximum potentially resolvable frequency is $f_{max} = 1/2\Delta t_{min}$, where Δt_{min} is the minimum sample interval, but also notes that '... the average value of Δt might better be used in defining a generalized Nyquist frequency – but which mean is appropriate: algebraic, geometrical, harmonic, ... ?'. From there, we can estimate lower and upper bounds for the effective Nyquist frequency as follows.

We can obtain a conservative, lower-bound estimate by calculating the arithmetic mean of the sample intervals, $\bar{\Delta t}_a$, and defining the average sampling frequency as $f_{s,a} = 1/\bar{\Delta t}_a$ and then the effective Nyquist frequency as $\hat{f}_{Nyq,l} = f_{s,a}/2 = 1/2\bar{\Delta t}_a$. Some software, including the R library package lomb, adopts this interpretation for its default calculation of the effective Nyquist frequency.

We can obtain a less conservative, upper-bound estimate by calculating the harmonic mean of the sample intervals, $\bar{\Delta t}_h$, and defining the average sampling frequency as $f_{s,h} = 1/\bar{\Delta t}_h$ and then the effective Nyquist frequency as $\hat{f}_{Nyq,u} = f_{s,h}/2 = 1/2\bar{\Delta t}_h$. This is equivalent to calculating the instantaneous sampling frequencies $f_s = 1/\Delta t$ and defining the effective Nyquist frequency as the arithmetic mean of the instantaneous sampling frequencies.

These are presented as estimates of the effective Nyquist frequency for unequal-interval data which should enable you to obtain meaningful 'full-spectrum' periodograms under many circumstances but which are not guaranteed to work under all circumstances. Under some circumstances, you will need to investigate the sampling in addition to the sampled data in order to estimate the effective Nyquist frequency. It is best to work with the more conservative, lower-bound estimate, at least in the initial stages of an investigation. If that doesn't reveal a frequency of interest, consider recalculating the arithmetic mean of the sample intervals with any statistical outliers excluded, or using the median rather than the arithmetic mean, provided these yield estimates within the upper-bound estimate.

Note that, in the limit as the unequal sample intervals converge to equal sample intervals, all these estimates (*e.g.* arithmetic mean, median, harmonic mean) for the Nyquist frequency converge to the true Nyquist frequency for equal-interval time-series.

9.2.3 Example: Exoplanet Orbital Period

The exoplanet_RV dataset in the library package astrodatR consists of time-series of radial velocity measurements for three stars with exoplanets. This dataset has four columns: column-1 is the identifier Star (HD.88133, HD.37124 and HD.3651), column-2 is MJD (time index, modified Julian days), column-3 is RV (velocity, ms^{-1}) and column-4 is S.D. (standard deviation, ms^{-1}). All three are unequal-interval time-series.

We will investigate star HD.3651 using the Lomb–Scargle periodogram command, lsp(), in the R library package lomb. First, we can extract the HD.3651 time-series as follows.

```
ex1 <- exoplanet_RV[exoplanet_RV$Star=="HD.3651",]
```

The default estimate of the effective Nyquist frequency can be calculated as follows, the first command calculates the sample intervals, the second the arithmetic mean and the third the effective Nyquist frequency.

```
dt <- diff(ex1$MJD)
dtam <- mean(dt)
fnyqd <- 1/(2*dtam)
```

We can then produce an initial Lomb–Scargle periodogram, as shown in Figure 9.5, using the default estimate of the effective Nyquist frequency, as follows (using default axis labels but suppressing the default title).

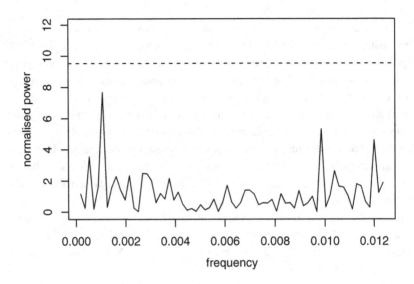

Figure 9.5 Basic Lomb–Scargle periodogram for HD.3651.

```
lsp(ex1$RV, times=ex1$MJD, main="")
```

In this case, the default estimate of the effective Nyquist frequency, *i.e.* $\hat{f}_{\text{Nyqd}} = 0.012$ cycles per day, is too low to reveal a frequency of interest at the (default) 1% significance level.

From there, we can investigate whether using a less conservative estimate of the effective Nyquist frequency reveals a frequency of interest. We estimate $\hat{f}_{\text{Nyq}}$ according to the harmonic mean and, for information, the median of the sample intervals as follows, using the sample intervals as calculated above.

```
dthm  <- 1/mean(1/dt)
fnyqh <- 1/(2*dthm)
dtmed <- median(dt)
fnyqm <- 1/(2*dtmed)
```

The first and second commands calculate the harmonic-mean-based estimate, $\hat{f}_{\text{Nyqh}} = 0.317$ cycles per day; the third and fourth commands calculate the median-based estimate, $\hat{f}_{\text{Nyqm}} = 0.170$ cycles per day. From there, we use the harmonic-mean-based estimate to produce the Lomb–Scargle periodogram (with axis labels and additional frequency-axis tick-marks and dotted vertical lines to indicate the orbital period, 62.23 days, and median-based estimate, $\hat{f}_{\text{Nyqm}}$, as annotated) as shown in Figure 9.6, as follows.

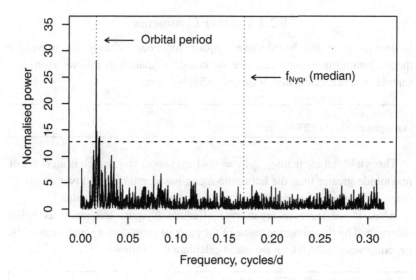

Figure 9.6 Refined Lomb–Scargle periodogram for HD.3651.

```
lsp(ex1$RV, times=ex1$MJD, from=0, to=fnyqh,
xlab="Frequency, cycles/d", ylab="Normalised power",
main="")
axis(1, c(1/62.23, fnyqm), labels=NA)
abline(v=c(1/62.23, fnyqm), lty=3)
text(0.1, 32, "Orbital period")
arrows(0.055, 32, 0.025, 32, length=0.05)
text(0.25, 25, expression(paste("f"[Nyq],", (median)")))
arrows(0.208, 25, 0.178, 25, length=0.05)
```

In Figure 9.6 there are two significant (at the default significance level) frequency markers, very close together, one visibly more significant than the other, at *ca.* 0.016 cycles per day, corresponding to the calculated orbital period of the exoplanet of 62.23 days, *i.e.* $f_{orb} = 1/62.23 = 0.016$ cycles per day (Fischer *et al.*, 2003). There is also a second pair of smaller (borderline) significant frequency markers, at *ca.* 0.02 cycles per day. In this example, using the median-based estimate of effective Nyquist frequency would have revealed the frequency of interest.

The frequency of interest is higher than the default maximum frequency, which is why we calculated the higher, less conservative estimate to produce this periodogram. In practice, with other data, it would be reasonable to expect to produce an initial periodogram using the default frequency interval before over-riding the default effective Nyquist frequency with a higher estimate.

9.2.4 Further Comments

If we were to regard the effective Nyquist frequency as being determined by the minimum sample interval, then we could calculate it as follows, using the sample intervals calculated in the preceding section.

```
dtmin <- min(dt)
fnyqmax <- 1/(2*dtmin)
```

This yields an estimate $\hat{f}_{\text{Nyqmax}} = 6.41$ cycles per day, which is an order of magnitude greater than the harmonic-mean-based estimate and two orders of magnitude greater than the default arithmetic-mean-based estimate.

Also, if we were to regard the effective Nyquist frequency as being determined by the minimum value of the pairs of consecutive sample intervals, *i.e.* equivalent to $2\Delta t$, then we could calculate it as follows.

```
dt2 <- diff(ex1$MJD, lag=2)
dtmin2 <- min(dt2)
fnyqmax2 <- 1/(dtmin2)
```

This yields an estimate $\hat{f}_{\text{Nyqmax2}} = 1.05$ cycles per day, which is approximately three times greater than the harmonic-mean-based estimate and an order of magnitude greater than the default arithmetic-mean-based estimate.

This illustrates the need the for caution when varying the maximum frequency of a Lomb–Scargle periodogram from the default value.

9.3 FFT Spectra and Significance

Having considered periodograms in outline and observed that we can establish criteria for considering probability and statistical significance, we will conclude with a consideration of the statistical significance of frequencies in the FFT spectrum.

Depending on the context, sometimes it is sufficient to note the presence of one or more frequency markers in a FFT (power) spectrum with magnitudes that are visibly several times larger than the background noise content, indicating a proportion of variance explained several times that of the background noise, but sometimes that is not sufficient which necessitates that we consider significance more formally. To do this, we will consider of the proportion of variance explained by individual FFT frequency components and then consider the F-ratio and F-statistics to obtain a p-value.

In considering the simple linear regression (or correlation) in Chapter 2, we have implicitly assumed null and alternative hypotheses which can be summarised as:

- null hypothesis, H_0, *i.e.* no statistical association between the time-series and the sinusoid we are regressing (or correlating) against greater than that we would expect due to noise;
- alternative hypothesis, H_1, *i.e.* significant statistical association between the time-series and the sinusoid we are regressing (or correlating) against greater than that we would expect due to noise.

The p-value is, paraphrasing, the probability that the null hypothesis applies and the result is termed statistically significant if the p-value is below a specified threshold, sometimes termed the significance level (sometimes denoted α). Thus, for a white-noise time-series, the expectation is that, with the exception of the zero-frequency marker, each frequency marker in the FFT spectrum has the same magnitude or, in practice, there are only insignificant differences in magnitude between markers in the FFT spectrum.

9.3.1 Power, Energy and Variance

If we were to systematically linearly regress a time-series against sinusoids with frequencies of $1, 2, 3, \ldots, M$ cycles over the duration, T, as introduced in Chapter 2, with M being the maximum resolvable frequency as defined in Chapter 4, then we would obtain the effect-size (*e.g.* coefficient of determination or gradient) and significance (*i.e.* p-value) for each frequency. However, this would be a laborious process – even if scripted on a modern fast CPU – and an unnecessary one because we can obtain the coefficients of determination, *i.e.* proportions of variance explained, directly from the FFT spectrum.

The power and energy in the time domain are directly related to the power and energy in the frequency domain by Parseval's theorem (M-A Parseval, 1755-1836), and we can use this to determine the proportion of variance explained by individual frequency components in the FFT spectrum. Stated in terms of time-series rather than signals that carry power and energy, Parseval's theorem states that the sum of the squares of the time samples is directly related to the sum of the squares of the magnitudes of the coefficients in the FFT spectrum. For an equal-interval N-point time-series and the standard form of the FFT (Equation 5.1), Parseval's theorem has the form

$$\sum_{n=0}^{N-1} x_n^2 = \frac{1}{N} \sum_{k=0}^{N-1} |X_k|^2 \tag{9.21}$$

where n and k are the time and frequency domain indices as we have used previously.

The variance of the time-series, σ^2, is given by

$$\sigma^2 = \frac{1}{N} \sum_{n=0}^{N-1} x_n^2 - \mu^2 \qquad (9.22)$$

where μ is the arithmetic mean of the time-series.

We can substitute from Equation 9.21 into Equation 9.22 to obtain

$$\sigma^2 = \frac{1}{N^2} \sum_{k=0}^{N-1} |X_k|^2 - \mu^2. \qquad (9.23)$$

From there, noting from Section 5.1 that X_0 is real and $|X_0| = N\mu$, we can rearrange the summation on the right-hand side and substitute to obtain

$$\sigma^2 = \frac{1}{N^2} \sum_{k=1}^{N-1} |X_k|^2 + \frac{1}{N^2} |X_0|^2 - \mu^2$$

$$= \frac{1}{N^2} \sum_{k=1}^{N-1} |X_k|^2 = \sum_{k=1}^{N-1} \left| \frac{X_k}{N} \right|^2. \qquad (9.24)$$

Thus, we have established that the variance of the time-series, on the left-hand side, is equal to the sum of the squares of the magnitudes of the normalised non-zero-frequency coefficients in the FFT spectrum, on the right-hand side. This confirms the observation in Chapter 5 that the power spectrum is often more informative than the amplitude spectrum because it is proportional to the variance spectrum.

9.3.2 Proportion of Variance Explained

In order to calculate the proportion of variance explained by a given frequency in a time-series, we divide both sides of Equation 9.24 by σ^2 such that kth term in the summation on the right-hand side becomes the proportion of variance explained by the kth component, for $k > 0$. However, it is more useful to re-index in terms of m, the order of the harmonic, $1 \leq m \leq M$ where M is the maximum frequency resolved by the FFT, as discussed in Section 4.7, to obtain the proportion of variance explained by the mth frequency sinusoid in the data. This yields

$$1 = \begin{cases} \dfrac{1}{\sigma^2 N^2} \displaystyle\sum_{m=1}^{M-1} \left(|X_m|^2 + |X_{-m}|^2 \right) + \dfrac{1}{\sigma^2 N^2} |X_M|^2 & N \text{ even} \\[12pt] \dfrac{1}{\sigma^2 N^2} \displaystyle\sum_{m=1}^{M} \left(|X_m|^2 + |X_{-m}|^2 \right) & N \text{ odd} \end{cases} \tag{9.25}$$

and, because the FFT frequency components are mutually orthogonal (linearly independent), each individual mth term on the right-hand side of Equation 9.25 is the proportion of the total variance explained by the mth frequency sinusoid, for $1 \le m \le M$.

This allows us to directly relate an individual mth term in Equation 9.25 to the coefficient of determination we would obtain if we were to explicitly linearly regress the time-series against that mth frequency sinusoid (or correlation coefficient if we were to correlate rather than linearly regress). Denoting the coefficient of determination for the mth frequency component as R_m^2 and extracting the mth term from Equation 9.25 , we obtain for $1 \le m \le M$

$$R_m^2 = \begin{cases} \dfrac{|X_m|^2 + |X_{-m}|^2}{\sigma^2 N^2} & \forall m, N \text{ odd}; \forall m \ne M, N \text{ even} \\[12pt] \dfrac{|X_M|^2}{\sigma^2 N^2} & m = M, N \text{ even.} \end{cases} \tag{9.26}$$

This is very useful: it means that we can directly obtain the proportion of variance explained for an individual m^{th} frequency sinusoid from the $\pm m^{\text{th}}$ coefficients in a FFT spectrum without having to explicitly synthesise that component from the FFT coefficients and then linearly regress (or correlate) the time-series against that synthesised component.

In fact, we can exploit the complex conjugate symmetry between X_m and X_{-m} to simplify this further to obtain (for real-valued data) for $1 \le m \le M$

$$R_m^2 = \begin{cases} \dfrac{2|X_m|^2}{\sigma^2 N^2}, & \forall m, N \text{ odd}; \forall m \ne M, N \text{ even} \\[12pt] \dfrac{|X_M|^2}{\sigma^2 N^2}, & m = M, N \text{ even.} \end{cases} \tag{9.27}$$

In practice, it is more straightforward than Equations 9.26 and 9.27 imply because the denominator in the terms on the right-hand side is the sum of the squared magnitudes of the non-zero-frequency coefficients in the FFT spectrum (Equation 9.24). Therefore, in practice, we do not need to calculate the variance of the time-series and all we need to do is normalise the non-zero-frequency power spectrum by dividing through by the sum of the non-zero-frequency powers.

To illustrate this, consider the Hillarys example in Section 5.3. We can obtain the variance spectrum by first deleting the zero-frequency row from the data-frame, then adding a new column for R^2, as follows.

```
hfop.t <- hfop.t[-1,]
hfop.t$Rsq <- hfop.t$pwr/sum(hfop.t$pwr)
```

9.3.3 Significance, p-value

For a white-noise time-series, length N and variance σ^2, *i.e.* its FFT spectrum has $N - 1$ non-zero-frequency components (positive and negative), our expectation is that each non-zero-frequency component in the FFT spectrum explains an equal proportion of the variance in the time-series, *i.e.* $\sigma^2/(N-1)$. The same basic principle applies in the cases of underlying red or blue noise but in those cases the expectation will depend on the precise frequency and the noise power–law.

From there, we can go one step further and estimate the statistical significance of the proportion of variance explained by a given frequency in the same manner as we would do if we were explicitly linearly regressing (or correlating) the time-series against the sinusoidal component at that frequency. If we were linearly regressing (or correlating) the time-series against a general mth sinusoidal component, we could calculate an F-statistic, as many programmed linear regression commands will do implicitly, and from there calculate a p-value.

The F-distribution, also known as the Snedecor, Fisher or Fisher–Snedecor distribution after G W Snedecor (1881–1974) who developed it from earlier work on analysis of variance by R A Fisher (1890–1962), is a continuous probability distribution which characterises the ratio of the explained variance to the unexplained variance. Summarising this in the context of linear regression with J explanatory variables of length N, we have the explained variance with J degrees of freedom and the unexplained variance with $N - (J + 1)$ degrees of freedom, and the F-statistic is

$$F = \frac{(tss-rss)/J}{rss/(N-(J+1))} \tag{9.28}$$

where rss and tss are the sum of the squared residuals and total sum of squares respectively, as in the definition of the coefficient of determination in Equation 2.8.

In terms of a coefficient of determination, R_j^2, for the proportion of variance explained by J explanatory variables, as in a multiple linear regression, we can re-express Equation 9.28 as

$$F = \left(\frac{R_j^2}{1 - R_j^2} \right) \left(\frac{N - (J + 1)}{J} \right) \tag{9.29}$$

and, from there, we can obtain the p-value as

$$p = 1 - \text{CDF}(F) \tag{9.30}$$

where $\text{CDF}(F)$ is the cumulative probability of the F-statistic for degrees of freedom J and $N - (J + 1)$.

In the immediate context, the J explanatory variables are orthogonal frequency components and, therefore, the total coefficient of determination is equal to the sum of the individual coefficients of determination, *i.e.*

$$R_j^2 = \sum_{j=1}^{J} R_j^2. \tag{9.31}$$

Therefore, we can calculate the F-statistic as follows expressing Equation 9.29 explicitly in terms of coefficients of determination (abbreviating the summation limits from Equation 9.31 for clarity)

$$F = \left(\frac{\sum_j R_j^2}{1 - \sum_j R_j^2} \right) \left(\frac{N - (J + 1)}{J} \right) \tag{9.32}$$

and, from there, we can obtain the p-value as in Equation 9.30.

In terms of a single general mth component with coefficient of determination R_m^2, noting $J = 1$, we have

$$F_m = \left(\frac{R_m^2}{1 - R_m^2} \right) (N - 2) \tag{9.33}$$

From there, we can obtain the p-value as in Equation 9.30, for degrees of freedom 1 and $N - 2$.

Non-uniform underlying (noise) spectrum

The foregoing is satisfactory when the time-series has an essentially white-noise characteristic, *i.e.* there is no frequency dependence in the underlying spectrum. However, if the underlying characteristic is not white noise, *i.e.* there is a frequency dependence in the underlying spectrum, then we might need to consider the variance in a frequency interval around the frequency of interest.

This requires that we consider Equation 9.27 and the subsequent formulae in terms of the variance in that frequency interval rather than the whole FFT spectrum.

9.3.4 Real Distinct-Frequency Data

In earlier chapters, we have used the nottem dataset and, informally, concluded that there are significant annual cycles in that dataset. Other than the frequency markers at 20 and 40 cycles over the 20-year period, there is no frequency dependence in the data, so we can follow the procedure outlined in the preceding section. To obtain the variance spectrum, in which each marker is the coefficient of determination for that frequency, we can extend the power-spectrum calculations of Section 5.2.4 (keeping the same steps and variable-names), as follows.

```
nottemf <- fft(as.numeric(nottem))
f.ind <- c(seq(0,240/2,1), seq(-(240/2-1),-1,1))
nottemfop <- data.frame(f.ind, nottemf)
nottemfop$f.cal <- nottemfop$f.ind/20

nottemfop$mag <- abs(nottemfop$nottemf)
nottemfop$pwr <- nottemfop$mag^2
nottemfop.t <- nottemfop[1:121,]
nottemfop.t$pwr[2:120] <- 2*nottemfop.t$pwr[2:120]
nottemfop.t <- nottemfop.t[-1,]
nottemfop.t$Rsq <- nottemfop.t$pwr/sum(nottemfop.t$pwr)
```

We can now extract the coefficients of determination and calculate F-statistics and p-values.

For the f_{20} annual sinusoid, we have

```
F.20 <- (nottemfop.t$Rsq[20]/(1 -
nottemfop.t$Rsq[20]))*(240-2)
p.20 <- pf(F.20, 1, 238, lower.tail=FALSE)
```

This yields $F.20 = 2486.2$ and $p.20 \approx 0$. Statistically, this is a highly significant result.

For the f_{40} six-month sinusoid, we have

```
F.40 <- (nottemfop.t$Rsq[40]/(1 -
nottemfop.t$Rsq[40]))*(240-2)
p.40 <- pf(F.40, 1, 238, lower.tail=FALSE)
```

This yields $F.40 = 3.72$ and $p.40 = 0.055$. Statistically, this result is slightly outside the *de facto* standard 5% threshold for 'standard' statistical significance.

For the combined annual-cycle, *i.e.* f_{20} and f_{40} sinusoids combined, we have

```
F.ann <- ((nottemfop.t$Rsq[20] + nottemfop.t$Rsq[40])/(1
- (nottemfop.t$Rsq[20] +
nottemfop.t$Rsq[40])))*(240-3)/2
p.ann <- pf(F.ann, 2, 237, lower.tail=FALSE)
```

This yields $F.ann = 1527.7$ and $p.ann \approx 0$. Statistically, this is a highly significant result: we conclude that this is a statistically significant annual cycle. In this case, the statistical significance clearly accords with our expectation but under other circumstances, where the periodic behaviours are not understood, you will need to interpret the statistical significance in the context of the data.

9.4 Summary

We have considered two periodogram approaches for analysing periodic content in unequal-interval time-series of two different types. It is important to remember that when using periodogram software, be cautious and considered if you need to vary the frequency limits from their default values, and check the documentation to make sure that you do not inadvertently exceed values considered safe or valid by the author of the software.

As well as giving us Fourier-related approaches for the analysis of unequal-interval time-series, consideration of periodograms has also allowed us to introduce concepts of statistical effect-size and significance by explicitly relating the power spectrum to the distribution of variance over the non-zero-frequencies. It is important to remember that this approach only considers statistical significance, in a manner consistent with correlation and linear regression. Importantly, it does not take account of the system from which a time-series is obtained. For example, the presence (*i.e.* magnitude greater than the background noise) or absence (*i.e.* magnitude less than the background noise) of a given frequency marker in a FFT spectrum might be significant of itself for the system under investigation. It is up to you as the investigator to assess the overall significance of the presence or absence of frequency markers in a FFT spectrum in the context of the data that you are investigating.

Exercises

1. Repeat the analysis of Section 9.2.3 to produce a Lomb–Scargle periodogram of the HD.3651 time-series with the median-based effective Nyquist frequency. Compare the significant frequencies to those revealed using the harmonic-mean-based effective Nyquist frequency.
2. As Fischer *et al.* do in their paper, the Lomb–Scargle periodogram can be plotted in terms of period rather than frequency using the type="period" option. Based on the information in Figure 9.6, choose sensible lower and upper limits and produce an HD.3651 periodogram calibrated in terms of period rather than frequency.
3. Using the same exoplanet_RV dataset, produce a Lomb–Scargle periodogram showing the orbital period of the exoplanet for star HD.37124 (154.46 days; Vogt *et al.*, 2005).
4. Extend the sunspots solar-variation investigation of Section 5.1 and calculate the proportion of variance explained by sinusoidal components with periods in the range $9.5 < T < 12.5$ years in the sunspot.year time-series and the statistical significance of the the combined components in this frequency range. Is it a reasonable conclusion that these arise by chance?

Appendix A   DFT Matrices and Symmetries

In Chapter 4 we have the basic form of an $N \times N$ DFT matrix, *i.e.*

$$
\begin{pmatrix}
1 & 1 & 1 & \cdots & 1 \\
1 & e^{-2\pi i\left(\frac{1}{N}\right)} & e^{-2\pi i\left(\frac{2}{N}\right)} & & e^{-2\pi i\left(\frac{N-1}{N}\right)} \\
1 & e^{-2\pi i\left(\frac{2}{N}\right)} & e^{-2\pi i\left(\frac{4}{N}\right)} & & e^{-2\pi i\left(\frac{2(N-1)}{N}\right)} \\
\vdots & & & \ddots & \vdots \\
1 & e^{-2\pi i\left(\frac{N-1}{N}\right)} & e^{-2\pi i\left(\frac{2(N-1)}{N}\right)} & \cdots & e^{-2\pi i\left(\frac{(N-1)^2}{N}\right)}
\end{pmatrix}
\tag{A.1}
$$

In this appendix, we explore some of the row (and column) symmetries in the coefficients and the implications of those symmetries on the frequency resolution of the DTF.

A.1 DFT Matrices, *N*-Even and *N*-Odd

There are key differences in the forms of the DFT matrix for *N*-even and *N*-odd. In the following two sub-sections, we will consider *N*-even and *N*-odd matrices respectively, simplifying the coefficients according to their 2π-periodic properties, *e.g.* $e^{-2\pi i\left(\frac{N-1}{N}\right)} = e^{-2\pi i\left(\frac{-1}{N}\right)} = e^{2\pi i\left(\frac{1}{N}\right)}$, to explicitly show the complex-conjugate row relationships.

A.1.1 *N*-Even

Taking the general form of the DFT matrix as in Equation A.1 and expressing it for general *N*-even, we have

159

$$
\begin{pmatrix}
1 & 1 & 1 & \cdots & 1 \\
1 & e^{-2\pi i\left(\frac{1}{N}\right)} & e^{-2\pi i\left(\frac{2}{N}\right)} & & e^{-2\pi i\left(\frac{N-1}{N}\right)} \\
1 & e^{-2\pi i\left(\frac{2}{N}\right)} & e^{-2\pi i\left(\frac{4}{N}\right)} & & e^{-2\pi i\left(\frac{2(N-1)}{N}\right)} \\
\vdots & & & & \vdots \\
1 & e^{-2\pi i\left(\frac{N}{2N}\right)} & e^{-2\pi i\left(\frac{2N}{2N}\right)} & & e^{-2\pi i\left(\frac{N(N-1)}{2N}\right)} \\
\vdots & & & & \vdots \\
1 & e^{-2\pi i\left(\frac{N-2}{N}\right)} & e^{-2\pi i\left(\frac{(N-2)\cdot 2}{N}\right)} & & e^{-2\pi i\left(\frac{(N-2)(N-1)}{N}\right)} \\
1 & e^{-2\pi i\left(\frac{N-1}{N}\right)} & e^{-2\pi i\left(\frac{(N-1)\cdot 2}{N}\right)} & \cdots & e^{-2\pi i\left(\frac{(N-1)^2}{N}\right)}
\end{pmatrix}.
$$

Simplifying this to explicitly show the complex-conjugate row relationships (ignoring column-wise simplifications for the time-being) we obtain

$$
\begin{pmatrix}
1 & 1 & 1 & \cdots & 1 \\
1 & e^{-2\pi i\left(\frac{1}{N}\right)} & e^{-2\pi i\left(\frac{2}{N}\right)} & & e^{-2\pi i\left(\frac{N-1}{N}\right)} \\
1 & e^{-2\pi i\left(\frac{2}{N}\right)} & e^{-2\pi i\left(\frac{4}{N}\right)} & & e^{-2\pi i\left(\frac{2(N-1)}{N}\right)} \\
\vdots & & & & \vdots \\
1 & e^{-\pi i} & e^{-2\pi i} & & e^{-(N-1)\pi i} \\
\vdots & & & & \vdots \\
1 & e^{2\pi i\left(\frac{2}{N}\right)} & e^{2\pi i\left(\frac{4}{N}\right)} & & e^{2\pi i\left(\frac{2(N-1)}{N}\right)} \\
1 & e^{2\pi i\left(\frac{1}{N}\right)} & e^{2\pi i\left(\frac{2}{N}\right)} & \cdots & e^{2\pi i\left(\frac{(N-1)}{N}\right)}
\end{pmatrix}. \tag{A.2}
$$

The first row in Equation A.2 is pure-real and clearly calculates the zero-frequency coefficient. However, note the form of the middle row in Equation A.2, $k = N/2$, *i.e.* frequency $f_{N/2}$: it is also pure real and simplifies to the sequence $1, -1, 1, \ldots, -1$ for all N-even. Thus, only cosine terms are 'seen' for this frequency (the Nyquist frequency).

A.1.2 *N*-Odd

Taking the general form of the DFT matrix as in Equation A.1 and expressing it for general N-odd, we have

$$
\begin{pmatrix}
1 & 1 & 1 & \cdots & 1 \\
1 & e^{-2\pi i\left(\frac{1}{N}\right)} & e^{-2\pi i\left(\frac{2}{N}\right)} & & e^{-2\pi i\left(\frac{N-1}{N}\right)} \\
1 & e^{-2\pi i\left(\frac{2}{N}\right)} & e^{-2\pi i\left(\frac{4}{N}\right)} & & e^{-2\pi i\left(\frac{2(N-1)}{N}\right)} \\
\vdots & & & & \vdots \\
1 & e^{-2\pi i\left(\frac{N-1}{2N}\right)} & e^{-2\pi i\left(\frac{2(N-1)}{2N}\right)} & & e^{-2\pi i\left(\frac{(N-1)(N-1)}{2N}\right)} \\
1 & e^{-2\pi i\left(\frac{N+1}{2N}\right)} & e^{-2\pi i\left(\frac{2(N+1)}{2N}\right)} & & e^{-2\pi i\left(\frac{(N+1)(N-1)}{2N}\right)} \\
\vdots & & & & \vdots \\
1 & e^{-2\pi i\left(\frac{N-2}{N}\right)} & e^{-2\pi i\left(\frac{(N-2)\cdot 2}{N}\right)} & & e^{-2\pi i\left(\frac{(N-2)(N-1)}{N}\right)} \\
1 & e^{-2\pi i\left(\frac{N-1}{N}\right)} & e^{-2\pi i\left(\frac{(N-1)\cdot 2}{N}\right)} & \cdots & e^{-2\pi i\left(\frac{(N-1)^2}{N}\right)}
\end{pmatrix}
$$

Simplifying this to explicitly show the complex-conjugate row relationships (ignoring column-wise simplifications for the time-being) we obtain

$$
\begin{pmatrix}
1 & 1 & 1 & \cdots & 1 \\
1 & e^{-2\pi i\left(\frac{1}{N}\right)} & e^{-2\pi i\left(\frac{2}{N}\right)} & & e^{-2\pi i\left(\frac{N-1}{N}\right)} \\
1 & e^{-2\pi i\left(\frac{2}{N}\right)} & e^{-2\pi i\left(\frac{4}{N}\right)} & & e^{-2\pi i\left(\frac{2(N-1)}{N}\right)} \\
\vdots & & & & \vdots \\
1 & e^{-\pi i\left(\frac{N-1}{N}\right)} & e^{-2\pi i\left(\frac{N-1}{N}\right)} & & e^{-(N-1)\pi i\left(\frac{N-1}{N}\right)} \\
1 & e^{\pi i\left(\frac{N-1}{N}\right)} & e^{2\pi i\left(\frac{N-1}{N}\right)} & & e^{(N-1)\pi i\left(\frac{N-1}{N}\right)} \\
\vdots & & & & \vdots \\
1 & e^{2\pi i\left(\frac{2}{N}\right)} & e^{2\pi i\left(\frac{4}{N}\right)} & & e^{2\pi i\left(\frac{2(N-1)}{N}\right)} \\
1 & e^{2\pi i\left(\frac{1}{N}\right)} & e^{2\pi i\left(\frac{2}{N}\right)} & \cdots & e^{2\pi i\left(\frac{N-1}{N}\right)}
\end{pmatrix}.
\qquad \text{(A.3)}
$$

The first row in Equation A.3 is pure-real and clearly calculates the zero-frequency coefficient. However, note that unlike the *N*-even form in Equation A.2, there is no single real middle row in Equation A.3 for *N*-odd. Instead we have a complex-conjugate pair for $k = (N\pm 1)/2$, *i.e.* frequencies at $f_{\pm(N-1)/2}$.

A.1.3 Specific *N*-Even and *N*-Odd DFT Matrices

On the following page, specific examples of *N*-even and *N*-odd DFT matrices are shown for $N = 10$ and $N = 11$ respectively to illustrate the general cases in the preceding two sub-sections. These values of *N* are big enough to explicitly show the symmetries in the real and imaginary parts of the matrix coefficients.

We can consider $N = 10$ as an example of an N-even DFT matrix, i.e.

$$\begin{pmatrix}
1 & 1 & 1 & 1 & 1 & 1 & 1 & 1 & 1 & 1 \\
1 & 0.81-0.59i & 0.31-0.95i & -0.31-0.95i & -0.81-0.59i & -1 & -0.81+0.59i & -0.31+0.95i & 0.31+0.95i & 0.81+0.59i \\
1 & 0.31-0.95i & -0.81-0.59i & -0.81+0.59i & 0.31+0.95i & 1 & 0.31-0.95i & -0.81-0.59i & -0.81+0.59i & 0.31+0.95i \\
1 & -0.31-0.95i & -0.81+0.59i & 0.81+0.59i & 0.31-0.95i & -1 & 0.31+0.95i & 0.81-0.59i & -0.81-0.59i & -0.31+0.95i \\
1 & -0.81-0.59i & 0.31+0.95i & 0.31-0.95i & -0.81+0.59i & 1 & -0.81-0.59i & 0.31+0.95i & 0.31-0.95i & -0.81+0.59i \\
1 & -1 & 1 & -1 & 1 & -1 & 1 & -1 & 1 & -1 \\
1 & -0.81+0.59i & 0.31-0.95i & 0.31+0.95i & -0.81-0.59i & 1 & -0.81+0.59i & 0.31-0.95i & 0.31+0.95i & -0.81-0.59i \\
1 & -0.31+0.95i & -0.81-0.59i & 0.81-0.59i & 0.31+0.95i & -1 & 0.31-0.95i & 0.81+0.59i & -0.81+0.59i & -0.31-0.95i \\
1 & 0.31+0.95i & -0.81+0.59i & -0.81-0.59i & 0.31-0.95i & 1 & 0.31+0.95i & -0.81+0.59i & -0.81-0.59i & 0.31-0.95i \\
1 & 0.81+0.59i & 0.31+0.95i & -0.31+0.95i & -0.81+0.59i & -1 & -0.81-0.59i & -0.31-0.95i & 0.31-0.95i & 0.81-0.59i
\end{pmatrix}$$

We can consider $N = 11$ as an example of an N-odd DFT matrix, *i.e.*

$$\begin{pmatrix}
1 & 1 & 1 & 1 & 1 & 1 & 1 & 1 & 1 & 1 & 1 \\
1 & 0.84-0.54i & 0.42-0.91i & -0.14-0.99i & -0.65-0.76i & -0.96-0.28i & -0.96+0.28i & -0.65+0.76i & -0.14+0.99i & 0.42+0.91i & 0.84+0.54i \\
1 & 0.42-0.91i & -0.65-0.76i & -0.96+0.28i & -0.14+0.99i & 0.84+0.54i & 0.84-0.54i & -0.14-0.99i & -0.96-0.28i & -0.65+0.76i & 0.42+0.91i \\
1 & -0.14-0.99i & -0.96+0.28i & 0.42+0.91i & 0.84-0.54i & -0.65-0.76i & -0.65+0.76i & 0.84+0.54i & 0.42-0.91i & -0.96-0.28i & -0.14+0.99i \\
1 & -0.65-0.76i & -0.14+0.99i & 0.84-0.54i & -0.96-0.28i & 0.42+0.91i & 0.42-0.91i & -0.96+0.28i & 0.84+0.54i & -0.14-0.99i & -0.65+0.76i \\
1 & -0.96-0.28i & 0.84+0.54i & -0.65-0.76i & 0.42+0.91i & -0.14-0.99i & -0.14+0.99i & 0.42-0.91i & -0.65+0.76i & 0.84-0.54i & -0.96+0.28i \\
1 & -0.96+0.28i & 0.84-0.54i & -0.65+0.76i & 0.42-0.91i & -0.14+0.99i & -0.14-0.99i & 0.42+0.91i & -0.65-0.76i & 0.84+0.54i & -0.96-0.28i \\
1 & -0.65+0.76i & -0.14-0.99i & 0.84+0.54i & -0.96+0.28i & 0.42-0.91i & 0.42+0.91i & -0.96-0.28i & 0.84-0.54i & -0.14+0.99i & -0.65-0.76i \\
1 & -0.14+0.99i & -0.96-0.28i & 0.42-0.91i & 0.84+0.54i & -0.65+0.76i & -0.65-0.76i & 0.84-0.54i & 0.42+0.91i & -0.96+0.28i & -0.14-0.99i \\
1 & 0.42+0.91i & -0.65+0.76i & -0.96-0.28i & -0.14-0.99i & 0.84-0.54i & 0.84+0.54i & -0.14+0.99i & -0.96+0.28i & -0.65-0.76i & 0.42-0.91i \\
1 & 0.84+0.54i & 0.42+0.91i & -0.14+0.99i & -0.65+0.76i & -0.96+0.28i & -0.96-0.28i & -0.65-0.76i & -0.14-0.99i & 0.42-0.91i & 0.84-0.54i
\end{pmatrix}$$

A.2 The Form of the DFT Frequency Spectrum

The preceding skeleton matrices for N-even and N-odd and the explicit $N = 10$ and $N = 11$ forms of the DFT matrix illustrate the basic symmetries in the coefficients along the rows. These symmetries also occur column-wise as well as row-wise. As we increase the number of rows (and columns), we increase the numbers of symmetries in the real and imaginary parts, whilst preserving key features such as:

- the complex-conjugate symmetries between rows 1 and $N - 1$, rows 2 and $N - 2$, rows 3 and $N - 3$, and so on towards the middle;
- the pure real middle row for N-even and the complex-conjugate middle rows for N-odd.

We can deduce that as row 0 clearly corresponds to the mean level (it yields N times the mean level), then row 1 corresponds to f_{+1}, row 2 corresponds to f_{+2}, and so on towards the middle. However, if row 1 corresponds to f_{+1}, then row $N - 1$ must correspond to f_{-1}, owing to the complex-conjugate symmetries. Similarly, if row 2 corresponds to f_{+2}, then row $N - 2$ must correspond to f_{-2}. Thus, we must have lowest frequencies in the rows at either end and the highest frequencies in the middle rows: for N-even, the highest frequency is given by a single row at $f_{N/2}$; for N-odd, the highest frequency is given by a complex-conjugate pair of rows at $f_{\pm(N-1)/2}$.

The Nyquist–Shannon sampling theorem (Section 4.8) formally proves that the maximum resolvable frequency for an equal-interval time-series (signal) is half the sampling frequency, $i.e.$ for an N-sample time-series, the maximum resolvable frequency is $f_{N/2}$, the Nyquist frequency. This is only achieved for N-even: for N-odd the maximum resolvable frequency is less than the Nyquist frequency by the factor $(N-1)/N$.

A.3 The FFT

Following on from the reasoning in the previous section, also note that we have essentially the same complex-conjugate symmetries column-wise as well as row-wise. Column 0 is pure real, there are complex-conjugate relationships between columns 1 and $N-1$, columns 2 and $N-2$, columns 3 and $N-3$, and so on towards the middle, and the considerations for N-even and N-odd for the middle columns correspond to those for the middle rows.

Thus, as well as pairwise symmetries for rows, there are pairwise symmetries for columns. All such symmetries enable us to reduce the number of calculations required to produce the DFT spectrum because we can replace some calculations with sign-changes, knowing that given combinations of rows and columns have to be complex-conjugate pairs.

This is the basis of the FFT: FFT algorithms exploit every possible symmetry to reduce the number of calculations for a given value of N, and produce all the other values by exploiting the row and column symmetries. As a general rule, the more factors a given value of N has, the higher the degree of optimisation of the FFT compared to the straightforward unoptimised DFT.

Appendix B   Simple Spectrogram Code

B.1 License

Like the parent software, R, this spectrogram code is free software and is provided
with ABSOLUTELY NO WARRANTY. You are welcome to modify and redistribute
it under certain conditions according to the GNU General Public License version 2 or
higher.

The code and help-file are available for download from www.cambridge.org/fourier.

B.2 The Program

B.2.1 How to Use the Code

The basic command, with options is:

```
simplespectro(signal, sput=1,
Nw=round(length(signal)/10), pstep=10, step=FALSE,
nlev=10, xlab="Time, time-units", ylab="Frequency,
cycles/time-unit", fmin=FALSE, fmax=FALSE, x0="0",
xT="T", Ntint=10, yextcal=FALSE, greyscale=FALSE)
```

and this produces a contour-plot time–frequency representation of a time-series or
signal.

If used with an output object on the left-hand side of the assignment operator (<-)
as follows:

```
sigspect <- simplespectro(...)
```

then **sigspect** is the matrix of STFT power spectra. Column-1 is the frequency
index and the other columns correspond to time-steps, as labelled; the rows correspond

to frequencies according to the range selected by fmin and fmax (as indicated in column-1).

The arguments are as follows:

- **signal** is the equal-interval time-series or, generally, equal-interval sequential data;
- **sput** is the number of samples per unit time, defaults to 1 but would, for example, be set to 12 for monthly sampled data where the base time-unit is years;
- **Nw** is the length (number of samples) of the STFT window, defaults to 10% of the length of the input time-series / signal / waveform, can't exceed half the length of **signal**;
- **pstep** is the step-length as a proportion (as a percentage between, but not including, 0% and 100%) of **Nw**, is overridden by setting **step=number**, for more precise control use **step**;
- **step** is the step-length (number of samples, integer), overrides **pstep**;
- **nlev** is the number of levels in the contour-plot, adjusted internally for sensible contours;
- **xlab** and **ylab** are the time (horizontal) axis label and frequency (vertical) axis label respectively;
- **fmin** and **fmax** are the minimum and maximum frequencies to plot, in cycles per unit time, time-unit as scaled by **sput**;
- **x0**, **xT** and **Ntint** are, respectively, the minimum and maximum time-axis labels and the number of intervals (tick-marks) between these;
- **yextcal**, setting **yextcal=TRUE** suppresses the internal default frequency axis, allowing the use of a custom-calibrated external frequency axis using **axis(2, ...)**;
- **greyscale**, the function defaults to the built-in rainbow colour palette, with red for maximum through to violet for minimum, setting **greyscale=TRUE** produces greyscale spectrogram plots as in Chapter 7.

The default options should produce a basic, unoptimised but informative spectrogram under many circumstances but this cannot be guaranteed and, in general, adjusting the settings will produce an optimised, more informative spectrogram.

B.2.2 The Code

This is the full code.

```
1   simplespectro <- function(signal, sput=1, Nw=round(
        length(signal)/10), pstep=10, step=FALSE, nlev=10,
        xlab="Time, time-units", ylab="Frequency, cycles/time
        -unit", fmin=FALSE, fmax=FALSE, x0="0", xT="T",
        Ntint=10, yextcal=FALSE, greyscale=FALSE){
2
3   m.sig <- as.matrix(signal)
4   dss <- dim(m.sig)
5
6   if(dss[2]>1 && dss[1]>1) stop("Exiting: not single
        column or row.")
7   if(dss[1]>1 && dss[2]==1) Ns <- dss[1]
```

```
8  | if (dss[2]>1 && dss[1]==1) Ns <- dss[2]
9  |
10 | ## trip illegals to defaults, not elegant but effective
11 |
12 | Nw <- round(Nw)
13 |
14 | if (Nw>(Ns/2))
15 | {
16 |     Nw <- round(Ns/10)
17 |     print("Invalid Nw: resetting Nw to default.")
18 | }
19 | if (sput <=0|sput >=Ns)
20 | {
21 |     sput <- 1
22 |     print("Invalid sput: resetting to default.")
23 | }
24 | else if(sput>=Nw) print("sput >= Nw: did you intend
   |     this?")
25 | if (step)
26 | {
27 |     step <- round(step)
28 |     if (step <=0 | step >=Nw)
29 |     {
30 |         step <- FALSE
31 |         pstep <- 10
32 |         print("Invalid step: resetting step & pstep to
   |             defaults.")
33 |     }
34 |     else print("Using entered step length, over-riding
   |         pstep")
35 | }
36 | if (pstep <=0|pstep >=100)
37 | {
38 |     if (!step)
39 |     {
40 |         pstep <- 10
41 |         print("Invalid pstep: resetting pstep to default
   |             .")
42 |     }
43 |     else print("Invalid pstep: overrident by step.")
44 | }
45 |
46 | ## window-step & final window start
47 |
48 | Nw1 <- Nw - 1
49 | fwi <- Ns-Nw+1
50 |
51 | if (!step)
52 | {
53 |     pstep <- pstep/100
54 |     step <- pstep*Nw
```

```
55          step <- floor(step)
56          if(step<1) step <- 1
57  }
58
59  ## window count and window start-indices, prettifying
        step
60
61  wc.frac <- (Ns - Nw)/step + 1
62  wc <- ceiling(wc.frac)
63  rwindex <- seq(1, fwi, length.out=wc)
64  windex <- round(rwindex)
65
66  ## STFT parameters
67
68  sg <- matrix(NA,Nw,wc)
69
70  for(k in 1:wc)
71  {
72      sg[,k] <- fft(signal[windex[k]:(windex[k]+Nw1)])/Nw
73  }
74
75  sg <- Mod(sg)
76  sg <- sg^2
77  if(Nw%%2==0)
78  {
79      Mw <- Nw/2
80      Mw1 <- Mw+1
81      f.ind <- 0:Mw
82      sg <- sg[1:Mw1,]
83      sg[2:Mw,] <- 2*sg[2:Mw,]
84  }
85  else
86  {
87      Mw <- (Nw-1)/2
88      Mw1 <- Mw+1
89      f.ind <- 0:Mw
90      sg <- sg[1:Mw1,]
91      sg[2:Mw,] <- 2*sg[2:Mw,]
92  }
93
94  f.cal <- f.ind*sput/Nw
95  f.cal.max <- max(f.cal)
96
97  ## plot preliminaries - frequency interval
98
99  if(fmin|fmax)
100 {
101     if(fmin)
102     {
103         if(fmin<0|fmin>=f.cal.max)
104         {
```

```
105                     rmin <- 1
106                     print("Invalid fmin: resetting to default.")
107             }
108             else
109             {
110                     rmin <- max(which(f.cal<=fmin)) + 1
111             }
112         }
113         else rmin <- 1
114
115         if(fmax)
116         {
117             if(fmax<=0|fmax>f.cal.max)
118             {
119                     rmax <- Mwl
120                     print("Invalid fmax: resetting to default.")
121             }
122             else
123             {
124                     rmax <- min(which(f.cal>=fmax))
125             }
126         }
127         else rmax <- Mwl
128
129         if(rmin>=rmax)
130         {
131             rmin <- 1
132             rmax <- Mwl
133             print("Invalid fmin - fmax combination:
                    resetting to defaults.")
134         }
135         else
136         {
137             sg <- sg[rmin:rmax,]
138             f.cal <- f.cal[rmin:rmax]
139         }
140 }
141
142 f.max <- max(f.cal)
143
144 ## plot preliminaries - decibels & scaling
145
146 sg <- sg/max(sg)
147 sg <- 10*log10(sg)
148 sglim <- range(sg)
149 levels <- pretty(sglim, nlev)
150
151 ## plotting
152
153 if(!greyscale)
154 {
```

```
155          filled.contour(1:wc, f.cal, t(sg), levels=levels,
               col = rev(rainbow(length(levels)-1)), xlab=xlab,
               ylab=ylab, plot.axes={axis(1, seq(1, wc, length.
               out=Ntint+1), labels=NA); axis(1, c(1,wc), c(x0,
               xT)); if(!yextcal){axis(2, NULL, NULL)}}, key.
               title = title(main = "dB", cex.main=1, font.main
               =1, line=0.5))
156      }
157      else
158      {
159          filled.contour(1:wc, f.cal, t(sg), levels=levels,
               col = rev(grey(0:length(levels)/length(levels))),
               xlab=xlab, ylab=ylab, plot.axes={axis(1, seq(1,
               wc, length.out=Ntint+1), labels=NA); axis(1, c(1,
               wc), c(x0,xT)); if(!yextcal){axis(2, NULL, NULL)
               }}, key.title = title(main = "dB", cex.main=1,
               font.main=1, line=0.5))
160      }
161
162      sg <- data.frame(sg)
163      names(sg) <- 1:wc
164      sg <- cbind(f.cal,sg)
165
166      invisible(sg)
167
168      }
```

B.3 Closing Comment

This is the version of **simplespectro** that was used to produce the figures in Chapter 7, and the time–frequency analysis to produce the cover image. If you want to explore 'spectrogramming' in more detail, with more versatile options, then look at some of the spectrogram options available for R in add-on library packages. For example:

- The **spectro** and **spectro3D** commands in the **seewave** package. Note that these are set up with a factor of 1000 between the time and frequency units, *e.g.* these calibrate frequencies in kHz for times in seconds.
- The **specgram** command in the **signal** package. This allows padding of the windowed sections to enable interpolation between frequencies over that allowed by the basic spectrogram window-length.
- The **spectrogram** command in the **phonTools** package. This assumes the input is a 'sound object': it is easiest to use the accompanying **makesound** command to convert your time-series into a pseudo 'sound object' and use that as the input to **spectrogram**.
- The **SPECT.drive** command in the **RSEIS** package.

All of these require more user-intervention than **simplespectro** to obtain a basic spectrogram, but allow various different options and output formats.

Further Reading and Online Resources

Further Reading

There are many books on Fourier theory and I cannot hope to list even all the introductory ones. This is a list of books that I have used over the years and have found useful – no more or less than that. My advice is to look around and find books that deal with aspects of the subject relevant to you in a manner that is accessible to you – and I hope that this book gives you a starting-point for accessing more advanced texts with confidence.

Dyke, P. (2014) *An Introduction to Laplace Transforms and Fourier Series* (2nd ed.), Springer Undergraduate Mathematics Series, Springer-Verlag London, ISBN:978-1-4471-6394-7.

Gabel, R. A. and Roberts, R. A. (1986) *Signal and Linear Systems* (3rd ed.), Wiley, ISBN:978-9-9715-1167-8.

James, J. F. (2011) *A Student's Guide to Fourier Transforms: With Applications in Physics and Engineering* (3rd ed.), Cambridge University Press, ISBN:978-0-5211-7683-5.

Kuo, F. (1966) *Network Analysis and Synthesis* (2nd ed.), Wiley, republished Wiley India Pvt. Limited, 2006, ISBN:978-8-1265-1001-6.

McQuarrie, D. A. (2003) *Mathematical Methods for Scientists and Engineers*, University Science Books, ISBN:978-1-8913-8929-7.

Newland, D. E. (1993) *An Introduction to Random Vibrations, Spectral & Wavelet Analysis* (3rd ed.), Longman, London, republished 2005 Dover, ISBN:978-0-4864-4274-7.

Proakis, J. G. and Manolakis, D. K. (2009) *Digital Signal Processing* (4th ed.), Pearson, ISBN:978-0-1322-8731-9.

Riley, K. F. (1974) *Mathematical Methods for the Physical Sciences: An Informal Treatment for Students of Physics and Engineering*, Cambridge University Press, ISBN:978-1-1391-6755-0.

Wirsching, P. H., Paez, T. L and Ortiz, K. (1995) *Random Vibrations: Theory and Practice*, John Wiley & Sons, New York, republished 2006 Dover, ISBN:978-0-4864-5015-5.

Online Resources

URL for the R software, documentation and manuals.
The R Project for Statistical Computing, www.r-project.org.

URLs for two comprehensive and accessible online resources I recommend to research
students. There are many others to choose from.
Bourke, P. (1993). DFT (Discrete Fourier Transform) FFT (Fast Fourier Transform),
http://paulbourke.net/miscellaneous/dft.
Shatkay, H. (1995). The Fourier Transform – A Primer. Technical Report.
Brown University, USA, ftp://ftp.cs.brown.edu/pub/techreports/95/cs95-37.pdf.
(Postscript versions also available.)

URLs for sources of data. These URLs were all live at the time of writing but it is likely
that some will change with the passage of time.
Tidal data are available from the Australian Baseline Sea-Level Monitoring Project
(ABSLMP), www.bom.gov.au/oceanography/projects/abslmp/data/index.shtml.
Regularly updated sunspots data are available from WDC-SILSO, Royal Observatory
of Belgium, www.sidc.be/silso/datafiles.
Global mean land-ocean temperature data are available from the NASA Goddard
Institute for Space Studies, http://data.giss.nasa.gov/gistemp/graphs.
Beryllium-10 solar-irradiance proxy data are available from the World Data Center for
Paleoclimatology, Boulder, and NOAA Paleoclimatology Program, ftp://ftp.ncdc
.noaa.gov/pub/data/paleo/icecore/greenland/summit/ngrip.
Climate data, e.g. Dubuque temperature data, are available from the Iowa State
University Mesonet, http://mesonet.agron.iastate.edu/climodat/index.phtml?station
=ia2364&report=16.
Mauna Loa atmospheric CO_2 data are available from the Carbon Dioxide Information
Analysis Center, http://cdiac.ess-dive.lbl.gov/ftp/maunaloa-co2.
Canadian atmospheric CO_2 data for Alert, Northwest Territories, are available from
the Carbon Dioxide Information Analysis Center, http://cdiac.ess-dive.lbl.gov/ftp/
trends/co2/altsio.co2.
The star data, and other datasets, are available from The University of Minnesota,
Duluth, www.d.umn.edu/~skatsev/Phys5053/Datasets.html.
USGS water resource data, including the Iowa river flow data, are available from the
National Water Information System, http://waterdata.usgs.gov/ia/nwis/sw.
Monthly Atmospheric and SST Index Values are available from the NOAA Climate
Prediction Center, www.cpc.ncep.noaa.gov/data/indices.
Various tree-rings and other datasets are available from the Time Series Data Library,
https://datamarket.com/data/list/?q=provider:tsdl.
Delta-^{18}O data are available from the NOAA National Centers for Environ-
mental Information, ftp://ftp.ncdc.noaa.gov/pub/data/paleo/contributions_by_
author/shackleton1990.
The exoplanet data are available from the Penn State University Center for Astrostat-
istics, http://astrostatistics.psu.edu/datasets/exoplanet_Doppler.html.

References

These are the books and papers specifically referenced in the text.

Dirac, P. A. M. (1930) *The Principles of Quantum Mechanics* (1st ed.), Oxford University Press.

Dirac, P. A. M. (1958) *The Principles of Quantum Mechanics* (4th ed.), Oxford at the Clarendon Press, ISBN:978-0-1985-2011-5.

Fischer, D. A., Butler, R. P., Marcy, G. W., Vogt, S. S. and Henry, G. W. (2003) A Sub-Saturn Mass Planet Orbiting HD 36511, *The Astrophysical Journal*, 590 (2), 1081–1087.

Fisher, D. A., Reeh, N., and Clausen, H. B. (1985) Stratigraphic noise in time series derived from ice cores. *Annals of Glaciology*, 7, 76–83, doi:10.3189/S026030550 0005942.

Heisenberg, W. K. (1927) Über den anschaulichen Inhalt der quantentheoretischen Kinematik und Mechanik, *Physik*, 43 (3,4), 172–198. doi:10.1007/BF01397 280. (German, English translation available at https://ntrs.nasa.gov/search.jsp? R=19840008978.)

Lomb, N. R. (1976) Least-squares frequency analysis of unequally spaced data, *Astrophysics and Space Science*, 39, 447–462, doi:10.1007/BF00648343.

Mann, M. E. and Lees, J. M. (1996) Robust estimation of background noise and signal detection in climatic time series, *Climatic Change*, 33, 409–445, doi:10.1007/BF 00142586.

Meyers, S. R. (2012) Seeing red in cyclic stratigraphy: Spectral noise estimation for astrochronology, *Paleoceanography*, 27, PA3228, doi:10.1029/2012PA002307.

Nyquist, H. (1928) Certain topics in telegraph transmission theory, *Trans. American Institute of Electrical Engineers*, 47 (2), 617–624, doi:10.1109/T-AIEE.1928.505 5024.

Lord Rayleigh. (1880) On the resultant of a large number of vibrations of the same pitch and of arbitrary phase. *Phil. Mag.*, Series 5, 10 (60), 73–78, doi:10.1080/14 786448008626893.

Scargle, J. D. (1982) Studies in astronomical time series analysis. II – Statistical aspects of spectral analysis of unevenly spaced data, *Astrophysical Journal*, Part 1, 263, 835–853, doi:10.1086/160554.

Schuster, A. (1897) On lunar and solar periodicities of earthquakes, *Proc. Royal Society of London*, 61 (369–377), 455–465, doi:10.1098/rspl.1897.0060.

Schuster, A. (1898) On the investigation of hidden periodicities with application to a supposed 26 day period of meteorological phenomena, *Terrestrial Magnetism*, 3 (1), 13–41, doi:10.1029/TM003i001p00013.

Shannon, C. E. (1949) Communication in the presence of noise, *Proc. IEEE*, 37, 10–21, doi:10.1109/JRPROC.1949.232969 & doi:10.1109/JPROC.1998.659497 (1998 republication).

Sloane, N. J. A. and Wyer, A. D. (Eds.) (1993) *Claude Elwood Shannon: Collected papers*, IEEE. ISBN:978-0-7803-0434-5.

Vaníček, P. (1969) Approximate spectral analysis by least-squares fit: Successive spectral analysis, *Astrophysics and Space Science*, 4 (4), 387–391, doi:10.1007/BF00651344.

Vaníček, P. (1971) Further development and properties of the spectral analysis by least-squares, *Astrophysics and Space Science*, 12 (1), 10–33, doi:10.1007/BF006 56134.

Vogt, S. S., Butler, R. P., Marcy, G. W., Fischer, D. A., Henry, G. W., Laughlin, G., Wright, J. T. and Johnson, J. A. (2005) Five new multicomponent planetary systems, *The Astrophysical Journal*, 632 (1), 638–658.

White, W. B., Lean, J., Cayan, D. R. and Dettinger, M. D. (1997) Response of global upper ocean temperature to changing solar irradiance, *J. Geophys. Res.*, 102 (C2), 3255–3266, doi:10.1029/96JC03549.

Index

aliasing, 97
  aliased frequency, 98
  aliased spectrum, 98
  foldback, 98
amplitude spectrum, 73, 93
autocorrelation, 125
  blue noise, 129
  red noise, 129, 134
  white noise, 128
autoregression, 125
  autoregressive model, 125
  blue noise, 127
  red noise, 126, 133
  white noise, 126

beryllium, ^{10}Be (data), 18, 29

camp (data), 136
circular functions, 7
co2 (data), 83, 90, 91, 93, 122
coefficient of determination, 24, 153
complex
  data, 44
  exponential, 42
  Fourier series, 42
  numbers, 2, 7
confidence interval, 27
correlation, 8, 10, 24, 27
  autocorrelation, 20
  coefficient, 10
  cross-correlation, 13
  lagged, 13
  negative, 10
  positive, 10
correlogram, 17, 18
  autocorrelogram, 20
covariance, 9

detrend, 85
DFT, 2, 56, 143
  derivation, 56
  forward, 59
  inverse, 61, 66
  linear transformation, 59, 60
  matrix, 59
    N-even, 159, 162
    N-odd, 160, 162
    symmetries, 159
  spectrum, 62, 163
    asymmetry, 63
Dirac delta-function, 51
Dirichlet conditions, 49
discrete Fourier transform, *see* DFT
dummy data, 26, 102

effect-size, 3, 44
EMD, 3
empirical mode decomposition, *see* EMD
Esperance (data), 90, 111
Euler's formulae, 42
exoplanet_RV (data), 147, 158

Fast Fourier Transform, *see* FFT
FFT, 2, 59, 163
  above-Nyquist features, 96
  adjacent frequencies, 107
  baseband spectrum, 97
  central spectrum, 97
  checking output, 71, 72
  Hamming window, 95
  Hann window, 95
  harmonic index, 62, 73, 79
  intermediate frequencies, 99, 100
  low-frequency features, 92

maximum frequency, 96
minimum frequency, 92
non-harmonic frequencies, 99
probability, 150
significance, 150
standard and non-standard, 69, 70
windowing, 94
discontinuity, 95
flow (data), 122
Fourier analysis, 1
Fourier series, 1, 6, 32, 41, 46
complex, 42, 56
Fourier theory, 1, 8
Fourier transform, 1, 47
integrability, 49
notation, 48
sinusoid, 51, 53
frequency, 34
angular, 36, 49
discrete, 45
function, 46, 48
fundamental, 25, 34, 45
harmonic, 25, 34, 45
index, 47
maximum, 63
negative, 43
Nyquist, 63–65, 146, 148
Nyquist rate, 65
F-statistic, 150, 154–156

hare (data), 30
HD.3651 (data), 147, 158
HD.37124 (data), 158
Hillarys (data), 78, 101, 105, 116, 122

index
frequency, 47
harmonic, 62, 73

lag, 14
least-squares spectral analysis, *see* LSSA
linear algebra, 7
linear combination, 1
linear dependence, 33
linear independence, 33
linear regression, 8
simple, 23, 24, 28
linearity
DFT, 61
Fourier series, 41
Fourier transform, 50

lomb (R library), 147
LSSA, 3
lynx (data), 30

Milankovitch cycles, 118
missing data, 109
linear interpolation, 110
mean value, 110
provisos, 110

nino3.4 (data), 132
noise, 87
blue noise, 125
Brown noise, 124
colour, 123
one-over-f noise, 124
pink noise, 124
red noise, 124
violet noise, 125
white noise, 124
Gaussian, 126

orthogonal, 33
orthogonality, 32
oxygen-18 (data), 118, 136

padding, 105
Parseval's theorem, 151
period, 34
periodogram, 7, 138
Lomb–Scargle, 4, 144
significance, 145
Schuster, 138, 143
significance, 141
phase, 39, 41
PortKembla (data), 90, 111
Portland (data), 111
power law, 124, 130
blue noise, 131
red noise, 130, 135
white noise, 130
power spectrum, 76, 80, 83, 87, 151
power spectral density, 124
red noise, 134
probability, 141, 145
p-value, 26, 27, 142, 146, 151, 154–156

R, 4
random walk, 138, 141
Rayleigh distribution, 142

re-windowing, 105
R-squared, *see* coefficient of determination

sampling theorem, 64
  Nyquist–Shannon, 64
Schuster's test, 141
short-time Fourier transform, *see* STFT
significance, 3, 26, 27, 154
singular-spectrum analysis,
      see SSA
spectrogram, 115, 169
  non-stationary data, 118
  resolution, 115
  simplespectro, 115
    code, 165
    how-to, 164
  stationary data, 116
spectrum, 45
SSA, 3
star (data), 111
stationarity, 112
  non-stationary, 112
  stationary, 112
  strong, 112
  weak, 113
statistics, 6
STFT, 113
  resolution, 114

sunspots (data), 15, 18, 20, 29, 87, 96, 122,
      158
symmetry, 36, 41
  even, 36
  Fourier series, 36
  Fourier transform, 50
  mixed, 39
  odd, 36

temperature
  land-ocean (data), 15
  Nottingham (data), 21, 25, 70, 90, 156
  tempdub (data), 90
tides
  diurnal, 79
  semi-diurnal, 79
  tidal harmonics, 82
treerings (data), 136
trend, 83

uncertainty principle, 114

variance
  explained, 152
variance spectrum, 151, 152
vector sum, 139, 141

wavelets, 3

Printed in the United States
By Bookmasters